D1082927

The Cave Bear Story

THE CAVE BEAR STORY

Life and Death of a Vanished Animal

Björn Kurtén

Columbia University Press

New York

Library of Congress Cataloging-in-Publication Data

Kurtén, Björn.
The cave bear story.

Bibliography: p.
Includes index.
1. Cave bears. 2. Paleontology—Pleistocene.
I. Title.
QE882.C15K88 569′.74 76-3723
ISBN 0-231-04017-2 ISBN 0-231-10361-1 (pbk.)

c 10 9 8 7 6 5 4 3
p 10 9 8 7 6 5 4 3 2 1

Columbia University Press
New York Chichester, West Sussex

Printed in the United States of America

They are all gone; sunk,—down, down, with thc tumult they made; and the rolling and the tramping of ever new generations passes over them; and they hear it not any more forever.

Thomas Carlyle
The French Revolution

Contents

Author's Note

No one escapes his fate. It might be said that my affair with the cave bear started half a century ago when it was decided to give the child a name that happens to be Swedish for bear. There were some early difficulties in living up to it, but in time it led to the distinction of a mention in the "Authors and Subjects" section of the *Journal of Insignificant Research.* Still, the real thing began in the early 1950s. Eager to apply newfangled population ideas on fossil mammals, I was casting about for a statistically respectable sample of some fossil mammal—any fossil mammal. The one that happened to be at hand had been collected a hundred years earlier, and had been lying, more or less forgotten, for many decades in the cupboards of the geology department of the University of Helsinki. It was the cave bear: hundreds and hundreds of teeth and bones.

I spent happy hours playing around with them and with the ensuing statistics. But, of course, there came the time when I had to go and find out if other cave bears, elsewhere, behaved the same way. Well, some did and some did not, and this made it necessary to look at other species of bears and see how *they* behaved. And there you are—there is no end to it.

The fossils told me a great deal about how the bears were put together and how they worked, but the best aspect of it all was that they revealed many other things as well. That is what I hope to show the reader in this book. The cave bear is its central figure; but the story is also about the excitement of discovery; the grandeur of the past; the dynamics of life and death in nature; evolution, its now well-traced course and rate; the relationship between bear and man; and the enigma of extinction. And, as I felt that the story could be interesting to many readers besides professional scientists, I have been careful to avoid unnecessary technicality.

In pursuing the cave bear, I have had the pleasure of meeting many distinguished colleagues who have followed the same trail. Among them are two of the leading cave bear authorities of this century: Professor Kurt Ehrenberg of Vienna and Dr. Frédéric-Edouard Koby of Basel. Between them, they disagreed strongly on many matters, such as Alpine dwarfing and the relationship between man and bear (I follow Ehrenberg in the former, and Koby in the latter). To me, they gave generously of their time, knowledge, and ideas. Ehrenberg also taught me to navigate the peculiar flat-bottomed skiff of his beloved Mondsee and introduced me to the beautiful Salzkammergut.

Koby, who died in 1969, was by profession an ophthalmologist, but he also became known as an amateur paleontologist and prehistorian of fully professional stature. With his steely eye and rich fund of sardonic humor, he was a relentless fighter in matters scientific, but personally he was the mildest of men. "He never killed an animal," writes Elisabeth Schmid, "not a fly, not a snake, not a mouse."

I wish to thank many other colleagues and friends for help, advice, and permission to see material in their care.

Among many others, I wish to thank K. D. Adam (Stuttgart), P. Boylan (Exeter), M. Crusafont (Sabadell), M. Degerbøl (Copenhagen), R. Dehm (Munich), C. C. Flerov (Moscow), A. D. Hallam (Taunton), D. A. Hooijer (Leiden), T. Lord (Settle), F. Prat (Bordeaux), D. E. Russell (Paris), R. J. G. Savage (Bristol), G. G. Simpson (Tucson), A. J. Sutcliffe (London), E. Thenius and H. Zapfe (Vienna), H. Tobien (Mainz), N. K. Vereshchagin (Leningrad), J. F. de Villalta (Barcelona), and E. Wegmann (Neuchâtel).

The original maps and anatomical illustrations for this book were drawn by Berit Lönnqvist. The life restorations of ancient animals and landscapes (except figure 2) were made by Margaret Lambert, and of these, figures 12, 18, 20 and 21 were drawn specially for this book. Dr. Wilfred T. Neill read an early draft of the text and made numerous helpful comments and corrections.

Helsinki, July 1975 Björn Kurtén

CHAPTER ONE

The Discovery

"The gruesome, terror-inspiring, immeasurable cavern, the lightless grotto, O Muse, bring back to my mind, and those death-like realms!

"We begin our walk, armed with walking sticks, ladders and torches. Stopping at the crest of the wood, we look for the lightless cave. . . . At long last it opens wide before our eyes, a cleft in the woods, in a desolate, distant area.

"I feel my limbs quaking as I step into the wide hall through the fissure in the rock. With silent steps I tread my way, and the cave widens and winding galleries lose themselves in deepest darkness, and in hundreds of corners and fissures. With torches I investigate the passages, and I see a thousand forms, which Nature has skillfully called forth, better than any human hand, from the living rock.

"Soon, as I look down, I see a greater number of horrendous human bones; teeth, too, I recognize; I can see bodies turned into stone, and skeletons left lying on the floor. . . ."

In these words Thomas Grebner, professor of theology and philosophy at the University of Würzburg, recorded his impression of a descent into the famous Zoolith Cave of Gailenreuth, in southern Germany, in the year 1748. This cave is one of several in the vicinity of the little town of Muggendorf, and they have been popular attractions for travelers for many years. Grebner, guided by a local gamekeeper by the name of Morsheuser, chose the most famous cave of them all for his descent, and it was the very cave that was to yield the first skull identified as belonging to the extinct cave bear.

What Grebner saw made such an impression on him that he wrote a description in the form of a poem in Latin hexameter: "Descriptio antri subterranei prope Galgenreuth," from which the excerpts above (in free translation) are taken. This poem is one of the earliest accounts of the observation of cave bear bones in a cave, though Grebner mistook them for human remains. It is true that some thirty years later another clergyman, F. J. Esper, did find some human bones in the cave earth of Gailenreuth, but the great mass of bones in the cavern certainly belonged to the cave bear.

In still earlier times, the bones of great creatures discovered in caves were thought to be those of mythical beasts: dragons, giants, or unicorns. Many a cave in Europe bears the name Dragon Cave or Dragon's Lair. The number of bones in some caverns is incredibly great, which of course excited curiosity, and in medieval and early postmedieval pharmacy such bones were pulverized and dispensed as curatives—mainly as an aphrodisiac. In addition, several "scientific" descriptions of dragon remains from caves were written; as late as 1672, J. Paterson Hayn identified bones from the Carpathian Mountains in this "scientific" manner. But by then there were scholars who recognized the bones for what they were—the remains of real animals, not mythical monsters. The enlightened attitude of these scholars was due to P. Bayer's *Oryctographica Norica,* printed in Nuremberg in 1632.

We may smile at Professor Grebner's "terror" and "quaking limbs," and there can be no doubt that a modern description of the Gailenreuth cave would read quite differently. And, of course, most of the "horrendous human bones" were actually bones of the cave bear. But we would do well to retain some of Grebner's enthusiasm and readiness to wonder and admire. For bones in caves, bones of those long dead, can still tell a story to those who, like Thomas Grebner, turn to the past in quest of adventure.

In his excellently illustrated treatise of 1774, Esper identified remains of bears in caves, compared them to those of the living brown bear, and commented that the main difference was that the cave form was much larger than the living. He also noted that certain teeth present in brown bears were absent in the fossil skulls. He thought the cave animal might have been a polar bear. But this theory was soon proved wrong by other scholars—among them the famous Darmstadt naturalist Johann Heinrich Merck and the equally renowned Scottish anatomist John Hunter. It remained, however, for Johann Christian Rosenmüller to give the animal a name.

Together with a friend, J. Chr. Heinroth, Rosenmüller published in 1794 the first description of the cave bear under the name *Ursus spelaeus* (meaning, literally, cave bear). Rosenmüller was then a young student, twenty-three years of age. The description was a thirty-four-page pamphlet in Latin entitled *Quaedam de Ossibus Fossilibus Animalis cuiusdam, Historiam eius et Cognitionem accuratiorem illustrantia,* which featured an engraving (by Rosenmüller) of a skull from Gailenreuth. This skull, then, is the type, or name-bearer, of the species, and the Gailenreuth cave is the type locality.

The type concept used to be taken very seriously by early systematists, and the specimen thus chosen was regarded as somehow embodying all the characters of the particular species. New finds were compared with it, and if slight differences could be observed, the new finds were regarded as members of another species.

Now it so happens that the cave bear skull, in common with those of other bears, is rather variable in shape. Moreover, with the immense material that proved to be available within a few decades, even unusual variants would turn up from time to time. In common with most other cave bears, for instance, the type skull has a markedly domed forehead (the reason for this doming will be discussed in chapter 2). But there are also skulls with a flatter profile, and these were regarded by Georges Cuvier, the leading French naturalist of the early nineteenth century, as members of a distinct species which he called *Ursus arctoideus* (because of their resemblance to the brown bear, *Ursus arctos*). However, Cuvier soon perceived his mistake, as he encountered a series of intermediate specimens connecting the two "types." He ended by firmly asserting the unity of the cave bear species.

Unfortunately the example of this great scientist was not followed by all, and scholars continued assiduously to erect new "species" of cave bears on the basis of what we now know to be simply individual, sexual or at most racial variants. There is a really surprising number of such superfluous species listed in the literature of systematics, and it will hardly serve any purpose to repeat them here (see notes at end of chapter).

In modern systematics, a "type" has a somewhat different meaning than it did in earlier centuries: it is now considered just a name-bearer. It does not have to be a

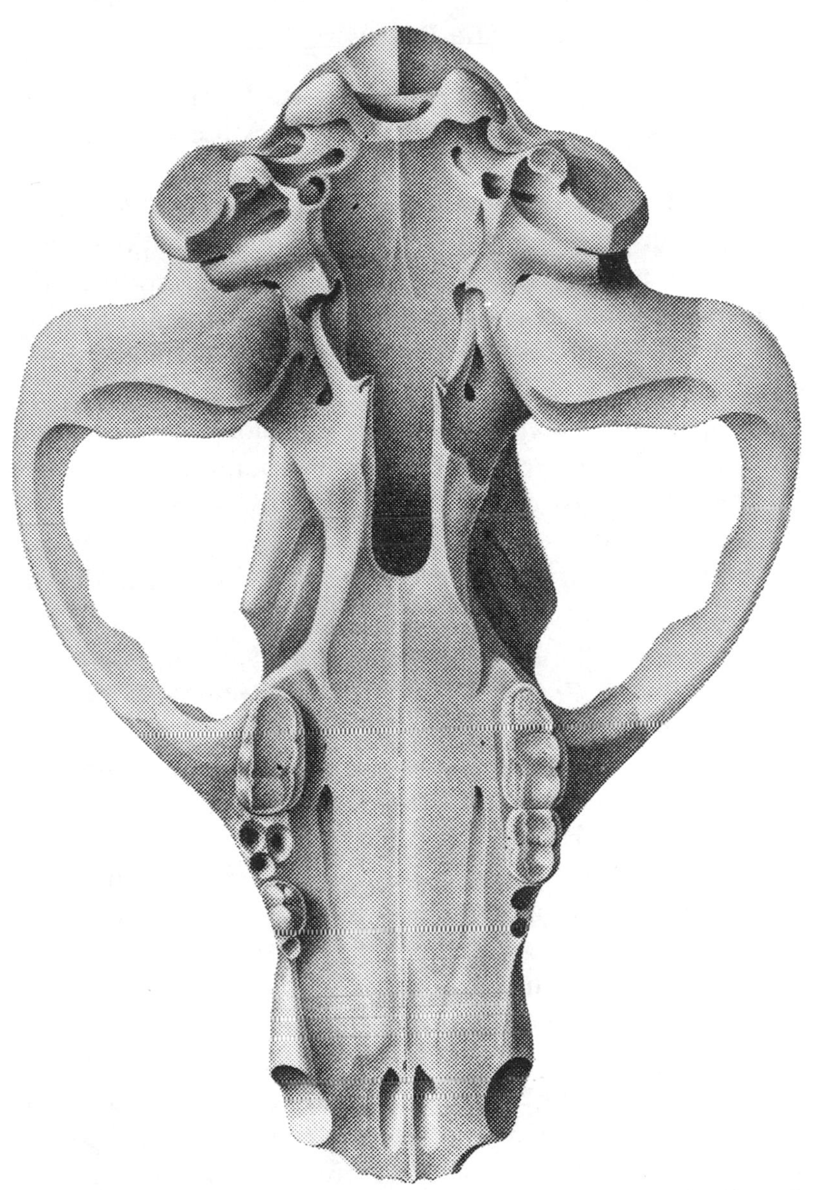

1. This picture of a cave bear skull, seen from below, was published by Alexander von Nordmann in 1858. Some of the teeth have fallen out, as shown by the empty root sockets. Normally, there are three cheek teeth (one premolar and two molars) on each side in the upper jaw, but in this specimen there is a small socket in front of the left premolar, showing that it retained a vestigial extra premolar. The skull belonged to a female individual, as shown by its relatively small dimensions and the small sockets for the canine teeth (eyeteeth).

"typical" or average member of its species. And when comparisons are made, the important thing is not the type, but the entire population to which it belongs.

As for the cave bear, though it has been dead and gone for thousands of years, there was soon quite a population at hand, a population of skulls and bones running into the hundreds, thousands, and more. In the nineteenth century bear caves with incredible numbers of fossil remains of this great animal were disclosed in Germany, France, Switzerland, Austria, Italy, Hungary, Czechoslovakia, Poland, Russia, and Belgium. Bear caves were found in Britain too, but the British cave bear turned out to be something quite special; we return to it in chapter 8.

Looking back at the early history of cave bear discovery, Rosenmüller stands out as its greatest figure and as a scientist of exceptional ability and perspicacity. Born in 1771 in Hessberg near Hildeshausen in Germany, he went as a student to the University of Erlangen. Here, he soon made himself a name as a speleologist, investigating the caves around the town of Muggendorf one of which now bears his name. He later became professor of anatomy at the University of Leipzig, rising to eminence in this field too; at least two organs of the human body are named for him, and his anatomy textbook appeared in new editions long after his death in 1820.

An early evolutionist, Rosenmüller (in a German version of the pamphlet describing *Ursus spelaeus*) discussed the possibility that changes in the environment, the food eaten, and so on, might result in the evolution of a new species. He noted that the cave bear might have become extinct because of climatic change, or alternatively, that the cave bear might have evolved into the living brown bear. It should be remembered that this was in 1795, long before the appearance of Jean Lamarck's evolutionary treatise of

1809 (as late as 1797, Lamarck expressly accepted the orthodox view that species were immutable).

Rosenmüller's scientific creed is expressed in the following words: "We should always, when forming opinions about the events in Nature, assume the most natural and common process; and if we endeavor to believe our senses rather than our imagination, we shall have no grounds for self-accusation."

Scientists, faced with the problem of these bear caves, had to find an explanation of how all those bones got into the caves. On the whole, there are four major hypotheses that have been brought forward from time to time.

The first was that the caves were formed in rocks that already contained fossil bones. The rock itself was dissolved, leaving the bones intact. This theory may sound fantastic and, in the case of the bear caves, it is so, but as a matter of fact it is not unusual to find petrified bone eroding out of a rock matrix. As far as cave bear fossils are concerned, however, it was shown that the limestone in which the caves were formed did not contain such bones; and we now know that the limestone was formed at the bottom of the sea many millions of years before the time of the cave bear. Indeed, this fact was pointed out by Rosenmüller in 1795.

Another theory, much espoused in the days when belief in a universal flood was still accepted as a basis for science, was that the bones were carried into the caves by running water, perhaps in the course of an enormous cataclysm. This is probably the reason why Esper thought the fossils might be polar bears, which live by the sea. But the nature of the caves and their sediments shows this idea to be impossible. It would indeed be a peculiarly selective flood that brought thousands of bear bones and little else into some caves, hyena bones into others, and so on.

The third theory was that the bear remains were

Chromotypographie SGAP

2. Impressed by the enormous numbers of cave bear remains, early scholars imagined that these animals formed great herds. In this old life restoration created by a French artist, men and cave bears are shown fighting in a subtropical landscape with monkeys clinging to the lianas of the trees.

brought into caves by man, who would then have been, in many cases, a specialized bear hunter. Although this suggestion was made as early as 1790 by H. Soemmering, the theory was not considered very seriously in the early days, for there was no good evidence that man had lived in the epoch of the cave bears and other extinct animals. Georges Cuvier, the founder of vertebrate paleontology and its leading authority in the early nineteenth century, thought that each geological epoch was terminated by a universal catastrophe in which the organisms then in existence were destroyed; this was followed by the creation of new plants and animals (or, alternatively, plants and animals immigrated from other areas that had not been destroyed). In Cuvier's scheme, the cave bear and its contemporaries belonged to an epoch before the creation of man. Discoveries of human bones in the same strata as those of cave bear, such as Esper's find from Gailenreuth, resulted from burials. Hence the phrase, "Fossil man does not exist," which has been attributed to Cuvier, although it may have originated with his followers who, as is usually the case, were more dogmatic than the founder of the theory.

The catastrophism of Cuvier gave way to uniformitarianism in the early 1830s when the British geologist Sir Charles Lyell showed that the geological processes in prehistoric times were the same as those that are still at work, changing the face of the earth in our own time; when Charles Darwin carried the theory of evolution to victory in 1859; and when evidence of the contemporaneity of early man and the extinct animals began to accumulate.

During many years of patient work in the ancient gravel terraces of the Somme River near Abbeville in northern France, the French archaeologist Jacques Boucher de Perthes found immense numbers of worked flints together with bones of animals long extinct. His discovery was not to be explained away, nor were MacEnery's and Pengelly's discoveries in Kent's Cavern, Torquay, England. There they found flint implements and weapons in deposits also containing the bones of bears and hyenas, and sealed on top by a thick, perfectly intact layer of hard stalagmite. There could have been no burial of the stone implements after the formation of the stalagmite.

Then, beginning in the 1850s, came the discoveries about primitive man himself, the Neandertaler, who had lived in Europe during the Ice Age, a contemporary of the cave bear, cave lion, mammoth, and woolly rhinoceros. The typical stonework of Neandertal man was identified and shown to be that of a big-game hunter. And so the theory that early man had been the agency that amassed the remains of bears and other animals in the Ice Age caves came to the forefront once more and is still influential.

The fourth theory concerning the mass occurrences of bears in the caves—and also hyenas and other animals—was that they came into the cave under their own power and died there, leaving their bones to be covered gradually by sediment. This opinion was put forward with excellent arguments by Rosenmüller. He evidently convinced most of his contemporaries, including Cuvier himself, and his explanation is still accepted by the great majority of scholars. Some years later, similar arguments were employed by William Buckland in England to account for the incredible amassing of hyena bones in Kirkdale Cave, Yorkshire. These arguments have only recently been challenged, but the problem of man and the bear, and man as a hunter of bears, is treated in chapter 6.

To facilitate modern studies of the cave bear and its world, we have many powerful new tools that were unknown to the pioneers. In the first place, there is now a solid body of geological information, which gives us a chronological framework into which to fit the age of the cave bear. Thanks to the new methods of radiometric dating,

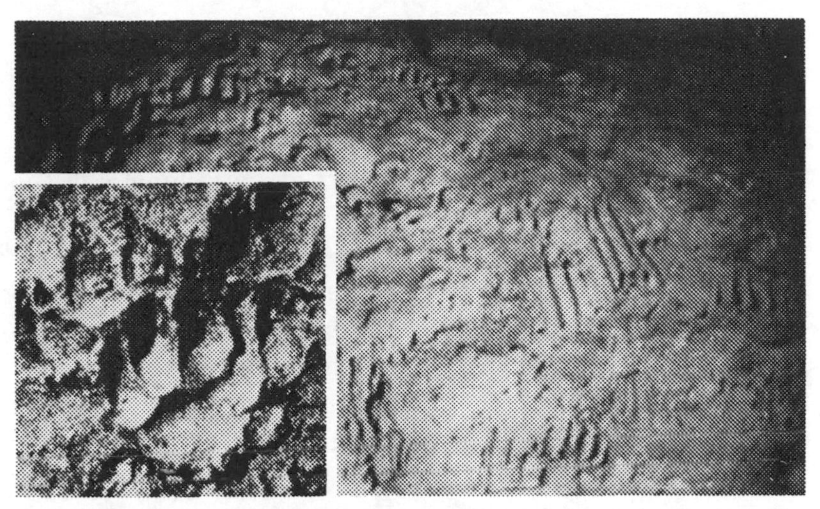

3. Cave bears have left footprints (inset) and scratch marks in the mud. The footprints are from Grotte Bijoux in the Pays Basque of southern France; the claw marks (photo M. Bouillon) are from Toirano Cave, Italy. After Kurtén.

which measure the decay of various radioactive elements, there is now a chronology with dates in years. The time scale of geology, now known to run into millions and even billions of years, would have been met with frank incredulity by most pre-Lyellian students. The ancestry of the cave bear, like that of all other living beings on earth, stretches back into those millions and billions of years, and we can trace at least part of it, as will be shown in chapter 3. But the time of the cave bear itself is much more recent. He lived, together with Neandertal men and later with men of our own species, during the latter part of the Ice Age and survived to or even beyond the end of that age.

The world of the Ice Age is being studied by a multitude of specialists. The geologist interprets the traces left by the inland ice. The paleozoologist identifies the remains of the animals of the past, the paleobotanist those of the plants. The archaeologist studies the prehistory of human cultures, the physical anthropologist the fossil remains of early man. The ecologist investigates the life histories of the various species and their interaction with each other and with the environment. The evolutionist traces the lineages of evolving populations through time. The paleoclimatologist interprets the facts in terms of past climates. Even the temperatures of ancient seas and lakes can be measured.

And so, in a way, the past comes to life again. Reindeer course through central Europe. Mammoths trumpet. The Irish elk spreads his immense antler shovels. The wild laughter of the cave hyena is heard once more. And the bear, deep in his dark cave, is awake now. It is time to make his acquaintance.

NOTES

Grebner's unpublished poem describing his visit to Gailenreuth was discovered by Heller (1956), who published a German translation. Esper (1774) identified the fossils as polar bears. The first descriptions of the cave bear as *Ursus spelaeus* were by Rosenmüller and Heinroth (1794) and Rosenmüller (1795). Other notable early treatises on the cave bear are by Cuvier (1823), Schmerling (1833), and von Nordmann (1858). A list of the various names bestowed upon fossil cave bears may be found in Erdbrink's comprehensive treatise (1953).

Lyell's classical work appeared in many editions, e.g. 1875. Buckland (1822) published on the mass occurrence of hyena in Kirkdale Cave.

CHAPTER TWO

Bear Bones

If you set up the skulls of a cave bear and a modern brown bear beside each other, you may note a sort of family likeness between them. Yet there are some differences too. If, in this chapter, we go into them in some detail, it is because both the similarities and differences are significant. They can tell us a great deal about the extinct cave bear and its mode of life.

In the first place, the cave bear skull is quite a bit larger than most modern bear skulls. It is true that there are a few present-day bears whose skulls reach about the same size as those of the largest cave bears. These bears are found along the coast of Alaska and British Columbia, on Kodiak Island and Afognak Island just south of Alaska, and seemingly also in a limited area of eastern Asia with a center in Manchuria. According to Boone and Crockett Club statistics, the world record skull of a brown bear has an overall length of 17.1 inches (456 millimeters) and a greatest width of 12.8 inches (325 millimeters); this prize specimen is in the Los

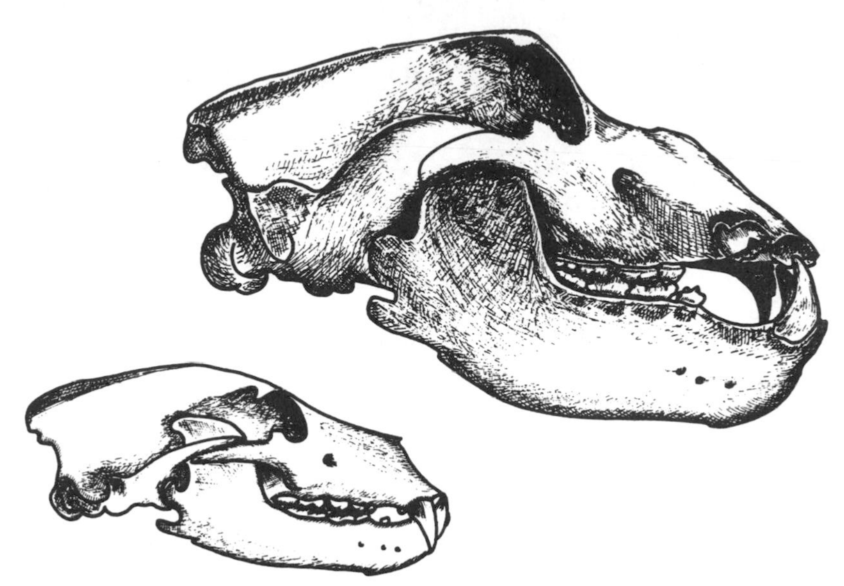

4. Skull of a male cave bear from a Swiss cave, compared with that of a modern brown bear from Finland, to the same scale. The right canine tooth of the cave bear was broken in life, and the bone around the stump bears witness to a heavy inflammation, which probably caused the death of the animal; the specimen is discussed further in chapter 7.

Angeles County Museum of Natural History. The skull of this Kodiak bear certainly rivals the largest cave bear skulls in size. But it is exceptional, and in general the brown bears of the Old World, and the grizzlies of the New, average distinctly smaller than the cave bear.

Another striking character seen in most cave bear skulls is a peculiar doming or "step" of the forehead, which is rounded just over the eyes and forms what is technically called a glabella. Not all cave bear skulls have this character fully developed, and there are even some in which the forehead is almost flat, but such variants are rare. The vaulted forehead gives the skull a subtly intellectual cast, which is, however, completely misleading, for what bulges inside is not brain but merely air-filled sinus cavities.

Are we to assume, then, that the cave bear skull is

domed in order to accommodate a set of very large sinuses? Not necessarily; there is another explanation, to which we return later. Before that, it should be mentioned that a step in the profile may also be seen in some brown bears, although never to the degree typical of the cave bears; the usual condition in the brown bear is a flatter, evenly curving forehead.

Looking once more at the bear heads in profile, we may note a distinct difference in the shape of the muzzle. Compared to the length of the jaw, the nasal bridge in the cave bear is quite short so that the nasal opening tends to face upward rather than forward. It looks a little like the flattened nose of a pug dog, and, in fact, many life restorations show the cave bear with a puglike face. I feel rather doubtful about this sort of restoration, for a retracted nasal bridge might also mean that the nose was well developed and movable. The nasal bones are actually very much shortened in animals provided with a trunk, like tapirs and elephants. Also, the cave bear does not have the underslung jaw of the pug. So I think the nose probably was not flat but quite long and protruding, and this is also the image seen in the art of Ice Age man, as we shall see in chapter 6.

Finally, still looking at the skulls in side view, note the lower border of the jaw. If you imagine these skulls resting on the top of a table, the skull of the brown bear would rest firmly on its straight lower jaw. That of the cave bear, in contrast, would be a veritable rocking chair. In a way this characteristic gives the final touch to the cave bear skull as a sort of curved and bulging thing, as if you had taken a cleanly proportioned brown bear skull and blown it up unevenly into a bizarre caricature. This curve-and-bulge character is by no means accidental. It has a distinct significance, and we can get to that by considering the purposes there are to a bear's head.

Speaking of purpose in science is called teleology. In

the old days, men assumed everything happened because of divine purpose. For modern science to develop, teleology had to be abolished. Instead of asking why, scientists asked how. When teleology returned to the life sciences, it had a very different meaning. The term should now be understood in terms of adaptation and natural selection. If an eye has a purpose—to see—it is because it has been fashioned for that function by adaptation, or that complex interplay between inheritance and environment, between internal chemistry and the outer world, that scientists term natural selection.

So the head of a bear, or of any mammal for that matter, has a purpose, or rather a number of purposes. To begin with, it serves to carry the main sense organs and the brain. Let us compare the eye sockets of the cave bear with those of the brown bear; we find that those of the cave bear are comparatively small, and we may conclude that its eyesight was relatively feeble. On the other hand, the nasal cavity of the cave bear is quite large and suggests a keen sense of smell. Dissection of the inner ear may give an idea of the acuteness of the hearing, but this has not been done.

The brain of the cave bear is not as large as you might expect from the size of the head. It is concealed far down at the back of the skull and is no larger than that of a normal brown bear. So it may be thought that the cave bear was not a particularly intelligent animal.

In addition to the two already mentioned, the head has another purpose: to grasp and chew food. This function is fulfilled by the jaws and teeth and the muscles that operate them. This apparatus is very large in relation to the other parts of the skull. To it belong not only the actual jaws and teeth, but also the great zygomatic arches that jut out on both sides, and the great crest that passes along the top of the skull, uniting at the back with an almost equally prominent cross crest. These arches and crests serve to attach the muscles that move the jaws.

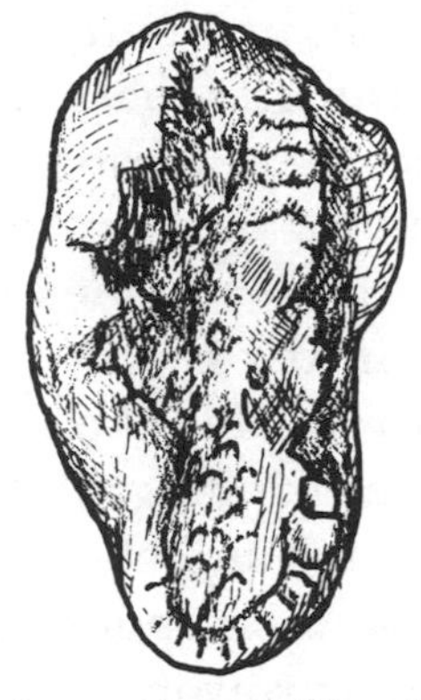

5. Molar teeth of a cave bear (left) and a wild hog, show resemblance in outline and cusp pattern, due to adaptation to a similar diet.

The actual business of grasping and chewing is performed by the teeth. Teeth can tell us a great deal about the mode of life of an animal. So let us take a detailed look at the teeth of the cave bear.

Bears are members of the order Carnivora, but most bears are far from exclusively carnivorous. They eat a wide range of animal and vegetable matter; they are omnivores, like man and the pig. And their omnivorous habits are reflected in the way that their teeth differ from those of other carnivores.

The typical carnivore set of teeth features small incisors in the front of the jaws for nipping and grasping; big eyeteeth or canines to worry and kill the prey; a set of sharp, meat-slicing premolars; and, in the back of the jaws, a set of many-cusped molars, which may be used to chew vegetable matter. Such a set of teeth is typical of a relatively unspecialized carnivore such as a dog. In the row of cheek teeth, the so-called carnassials stand out clearly; they are two great slicing teeth, the hindmost upper premolar and the foremost lower molar, which bite against each other. You can see a cat or dog use these teeth when it chews meat.

In extreme flesh eaters like the cats, the blunt-cusped molars behind the carnassial teeth tend to dwindle and vanish. This deterioration can be followed in the fossil record. On the other hand, in those carnivores that eat a great deal of vegetable food—such as bears and badgers—the back teeth tend to become larger and to develop broad masticatory surfaces with numerous low, rounded cusps. This trend is carried to an extreme in the bears, with the result that their teeth have come to resemble those of pigs to a surprising degree. In bears the grinding molars are very large, the carnassials relatively small and blunt, and the premolars much reduced. As the sole exception among living bears, the polar bear is strictly carnivorous, and its teeth have responded. Although still characteristic bear teeth in other respects, their originally rounded cusps have become high and pointed.

If the bear trend towards vegetarianism was sidetracked in the polar bear, it seems to have gone on to a veritable culmination in the cave bear. The French paleontologist Albert Gaudry called the cave bear "le moins carnivore des Carnivores et le plus ours des Ours"—the least carnivorous of carnivores, and the most bearish of bears.

The cave bear molars are much lengthened and expanded, and it is clear that they had tough work to do, grinding whatever food it was that the cave bear lived on. In old bears teeth wear down to senile stumps; the entire crown may eventually wear away, and in the end the roots are wearing down too. You can see this condition occasionally in very old brown or grizzly bears; in skulls of cave bears it is a common sight.

Such a state is almost certainly the result of eating vegetable matter, for plant cells need much more chewing than animal cells. The plant cell is encased in a tough cellulose skin. We ourselves, when eating vegetables, break down the cell walls by boiling, roasting, or baking, or perhaps by grating the raw vegetable (which amounts to a kind of preliminary chewing) so we do not have to wear out our teeth in the process of mastication. Such tricks are beyond the bear, who is entirely dependent on his big grinders. The same, of course, was true for early man, before he invented the use of fire, and his molars were up to twice the size of ours.

In front of the set of grinders (which includes the hindmost, or fourth premolars), toothless gums extend forward in the cave bear to the great canine teeth or eyeteeth, and the well-developed row of incisors. Such a toothless interval is termed a diastema, and the diastema in the cave bear marks the place originally occupied by the three anterior premolars. In brown bears, some of the anterior premolars, though rarely all three, are still present as small peglike structures; but in the cave bear, with rare exceptions, they are all gone. Compare this situation with that of the dog, in which the premolars are sharp, slicing teeth, or the hyena, in which they have become powerful bone-smashing structures. Clearly, the bears are on a very different evolutionary tack, and again we find the cave bear as the culmination, the "most bearish of bears."

Looking at the cave bear's dental battery, we can see that the main part of the action took place relatively far back in the tooth row, in the big grinders. Now, for chewing, all mammals need not only jaws and teeth, but also muscles. There must be muscles to open and close the jaws and to move them against each other in a grinding manner.

Not much force is needed to open the mouth, for the lower jaw will tend to fall open of its own weight, which can happen to humans when our muscles relax under the impact of a great surprise. Accordingly, there is a comparatively weak muscle, the digastric, extending from the lower border of the jaw backward to the base of the skull. In its contraction, the digastric pulls the jaw down and back.

Closing the jaws, on the other hand, requires great strength, and this function is handled by two powerful muscle complexes. One is the masseter muscle, extending from

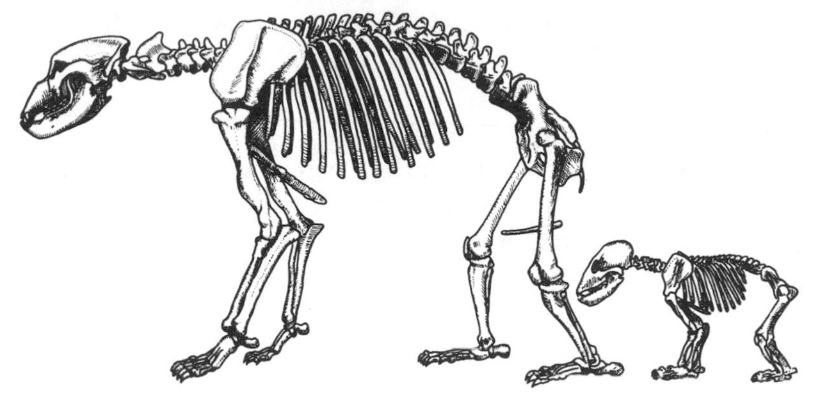

6. Skeletons of an adult male cave bear and a seven-month cub. The adult skeleton is composite; the cub (after Ehrenberg) was found in an Austrian cave.

the jawbone upwards and outwards, inserting into broad zygomatic arches—the cheekbones. Because of its direction obliquely upward and outward, the masseter not only closes the jaw but also can pull the jaw from side to side, giving it a grinding motion.

The other muscle involved in chewing is the temporal muscle that stretches from the jawbone, and especially from its ascending part—the coronoid process—to the top and back of the head, as a mighty sheet of muscle fibers. It is this muscle that fashions the crest along the top of the skull, into which the jaw is anchored, and the muscle also inserts into the cross crest farther back.

The development and needs of the masseter and the temporal muscles strongly influence the architecture of the skull. If a strong grinding motion from side to side is required, the zygomatic arches have to jut broadly to the side, as they do in the bear. As to the temporal muscle, its development is affected by the part of the dentition that takes the greatest strain. If the strain is in the front part of the jaws, the temporal muscle fibers will be almost horizontal, and the main pull of the muscle goes from the coronoid process to the back of the skull. Such is the case in cats and hyenas, to take two instances among carnivores, or in the apes, to take an example among the primates. If, on the other hand, the main business of the dentition is enacted at the back of the jaws, then the muscle fibers have to be more nearly vertical and the main direction of pull is towards the crown of the head. Such is the case in the cave bear—and in ourselves, for that matter. And here we come to the reason for the peculiar profile of the cave bear.

A muscle needs length. If an animal has a long, low head and needs vertical chewing muscles, the only way they can be long enough is for the crown of the head to be raised. In man this presents no problem; the great expansion of his brain gives him a lofty skull, far higher than anything needed by his temporal muscles. In the bear it is different. Its brain is much too small to fill out the expanding skull, and solid bone would be too heavy; so the skull is filled with empty sinus cavities. While the crown of the head is raised, there is no need to raise the top of the muzzle. Thus the cave bear has a stepped profile with a low nasal profile and a high glabella.

So there is a direct relationship between the development of the big grinders and the vaulting of the forehead in the cave bear. In the brown bears, where the grinders are smaller, the profile is flat or only moderately stepped. And the polar bear, which has still weaker back teeth but quite powerful carnassials, has a very long, low, almost tubular skull with a perfectly flat forehead.

In a meat-eating carnivore like the cat, the cheek teeth of the upper and lower jaws move past one another like the blades of a pair of scissors. If you draw a line along the bases of the cheek teeth, you will find the pivot of the jaw joint in the line's elongation. But if you draw a similar line for the cave bear, you will find that the jaw joint is raised well above the line, and this is the condition that induces the strong curvature of the lower border of the jaw (see figure 7).

In a scissor motion, the two blades act against each

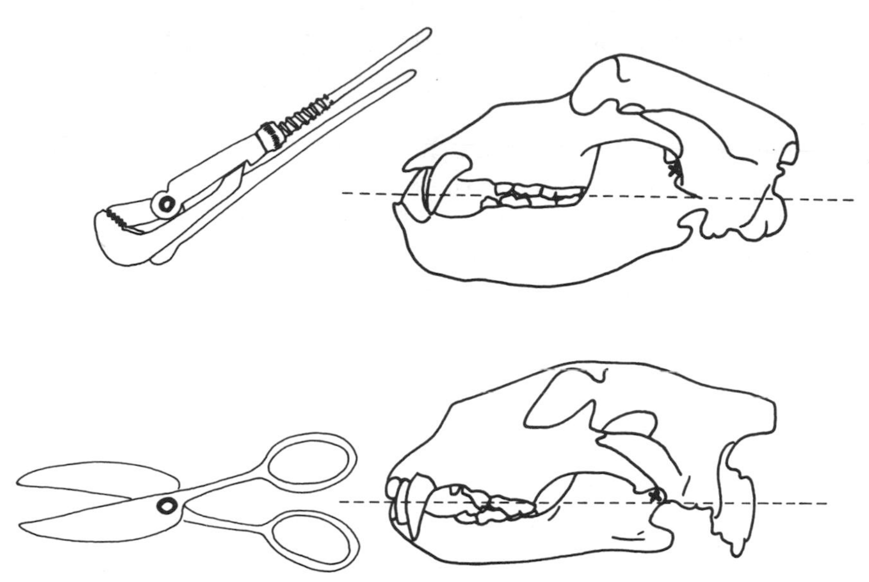

7. Jaw mechanics of the cave bear may be compared to that of a spanner, with the pivot of the jaw joint (cross) raised above the occlusal plane of the cheek teeth. In contrast, the pivot of the cat skull (as exemplified by a lion, below) is in the same plane as the teeth and the action resembles that of a pair of scissors.

other only at one point at a time, and this point moves forward as the scissors close. But if a nutcrackerlike action is needed, in which the whole set of teeth acts at the same time, the pivot must be located well above or below the level of the teeth. This is the type of action suitable to omnivores and especially to vegetarians, and it is found not only in the cave bear but also in all the hoofed grazing animals—horses, antelope, buffalo, etc. Not to mention ourselves!

So we see that the peculiarities of the cave bear skull are mostly bound up with its feeding habits, and that far from being haphazard, they are strongly adaptive and have been brought about by natural selection.

After this detailed look at the skull of the cave bear, the remainder of its body may be considered more briefly. Most readers will have seen a brown or grizzly bear, in a zoo if not in the wild, and retain a memory of that thickset, extremely powerful, yet surprisingly agile animal. While attaining the same size as the largest grizzlies or brown bears, the cave bear was rather differently proportioned. Its length from nose to base of tail was over five feet (1.6 meters), its height about four feet (1.2 meters). These figures would represent an average male; the female was noticeably smaller.

If we think an ordinary bear heavy of build, the cave bear with its barrellike body carried this trait to the extreme. The large head was borne on a moderately long neck, apparently very often in a low-slung position near the ground. The limbs were rather short but very powerful, with broad, short feet turning inward even more sharply than in the brown and grizzly bears. The feet were armed with powerful claws, shorter than those of the grizzly but very stout, which may have been used for digging as well as for attack or defense. All of the limb and foot bones of the cave bear are much stouter and heavier than those of the brown bear, and thus are easy to identify.

There is even direct evidence of the tremendous breadth of the cave bear paw, in the shape of scratch marks made in the soft loam of the cave fill or on the floors and walls of various caves. In some cases these marks may have been made when the bear slipped on the moist surface. There are usually three to five parallel scratches, and the distance between the outermost marks may be up to 5.5 inches (14 centimeters). In a modern brown bear, the corresponding figure is about 3.9 inches (10 centimeters).

After observing the limb proportions of the cave bear, the image of a slow, rolling gait grows stronger. In four-legged animals there are three major segments to each limb. For instance, the hind limb is comprised of the thigh, the shin, and the foot. Their length relationships give good clues to the manner in which a particular animal moved. A short thigh and a long shin means short leverage for the

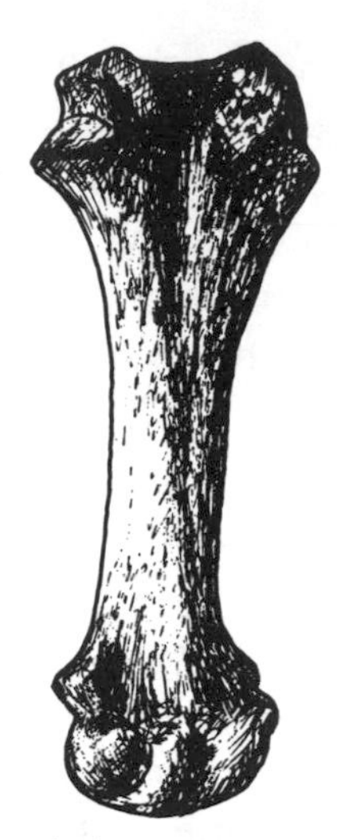

8. Shortness and plumpness distinguish the cave bear foot bone (right) when compared to that of the brown bear. These are first metatarsals, or middlefoot bones of the inner toe in the hind foot. The cave bear bone comes from Odessa, USSR; the brown bear bone belonged to an animal that lived during the penultimate glaciation (the Saalian) in Devon, England, and was found in Tornewton Cave.

muscles that move the limb forward and backward, and consequently a rapid but not very powerful stride. Such construction points to a lightly built, speedy creature, such as the fox. In contrast, elongation of the thigh and shortening of the shin and foot means power but also slowness, and so presumably a heavy, slow-moving animal.

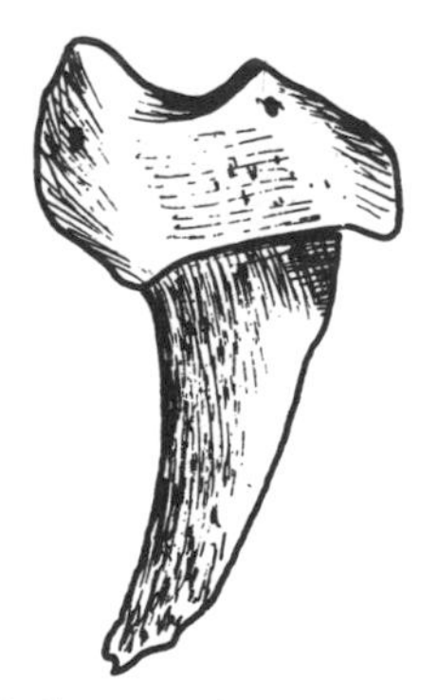

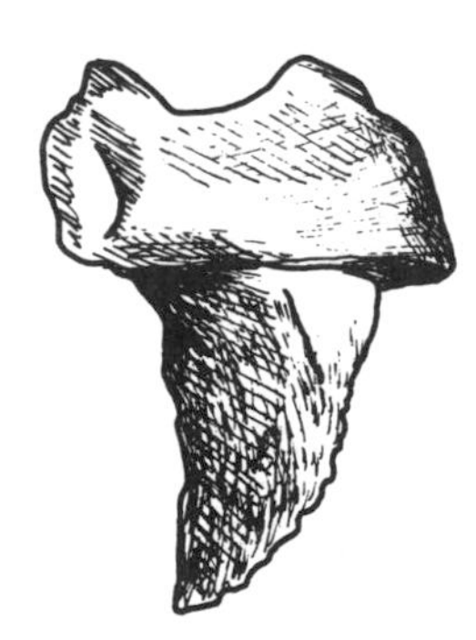

9. Claw phalanges of a cave bear and a modern brown bear. The relative shortness and the heavier build of the cave bear are again evident.

It is no surprise, then, to discover that the cave bear indeed had a much longer thigh, but a shorter shin and foot, than the brown or grizzly bear. Much the same holds for the front limb, in which the humerus (upper arm bone) is very long but the forearm and hand comparatively short.

How much did the cave bear weigh? This can be estimated and in various ways. One good method would be to make a careful life restoration of the bear in the form of a sculpture. (There is an excellent one in the natural history museum in Basel, Switzerland.) The volume of the sculpture would then be measured by immersing it in water. To find out the weight of the cave bear it is necessary also to know the specific gravity of a living bear. To my knowledge this has not been done for bears, but a somewhat similar study was carried out on crocodiles and dinosaur restorations some years ago in an attempt to estimate the weights of some of the extinct dinosaurs.

The cave bear's weight can also be estimated directly from some of the bones of its skeleton. For example, the limb bones have to withstand the weight of the animal and have to increase in thickness as the weight grows greater. A comparison of the cross section of the femur (thighbone) of a cave bear with that of a brown bear reveals that the cross-sectional surface for a male cave bear is about 2½ to 3 times greater than for a male European brown bear. The latter weighs about 350 pounds (160 kilograms) on the average, so the weight of the cave bear should then be about 900 to 1,000 pounds (400 to 450 kilograms). These estimates relate to large males and to fairly lean individuals, not to the superfatted bear seeking its den in late autumn, when it probably weighed a great deal more. On the other hand, the smaller female cave bear probably reached little more than half the weight of the male.

Many bears, especially the smaller species but also the young of the larger, are good climbers and often take to the trees. The weight and build of the cave bear suggests that it could not have been much of a tree climber, and this impression is verified by analysis of the shoulder blade. Its shape shows that the muscles used in climbing were relatively feeble, while those used in walking and digging were strong.

The image of the cave bear now emerges very clearly. We can visualize it as a great, ponderously lumbering bear moving about with its muzzle close to the ground most of the time. Although not a tree climber (except perhaps the cubs), it would move easily in alpine terrain. Its food would probably consist of succulent plants, berries, roots and tubers dug out of the ground, tender grass, small animals, and so on, and it would certainly not disdain the carcases of prey brought down by doughtier hunters such as the contemporary cave lion, leopard, or cave hyena.

10. Restoration of a cave bear by Margaret Lambert. After Kurtén.

NOTES

The Mixnitz monograph edited by Abel and Kyrle (1931) contains a wealth of information on cave bear anatomy. See also Ehrenberg (1931, 1935a, b, 1942, 1964, 1966), Erdbrink (1953), and Mottl (1933). Dinosaur weights have been estimated by Colbert (1962). Cave bear weights are estimated in Kurtén (1967a).

CHAPTER THREE

Origins

Few extinct animals are known from such a great number of fossil remains as the cave bear. Because there is such a wealth of information, a uniquely detailed picture of its anatomy and life history can be constructed. The same may be said of the origin of the cave bear. Almost every stage in its history can be traced back, in unbroken lineage, for 5 million years or more. And we occasionally get glimpses of still older stages in this long history.

Our story may well begin about 20 million years ago, in the early part of the Miocene epoch of earth history. The place is what is now called Wintershof-West in the Bavarian mountains of southern Germany. In those very distant times, Europe was a subtropical land. Moisture-laden monsoons blew in from an Atlantic that was narrower and a Mediterranean that was broader than their counterparts today. Much of the continent was clothed in luxuriant forests in which grew palms, camphor trees, and many other warmth-loving species. The rivers teemed with croco-

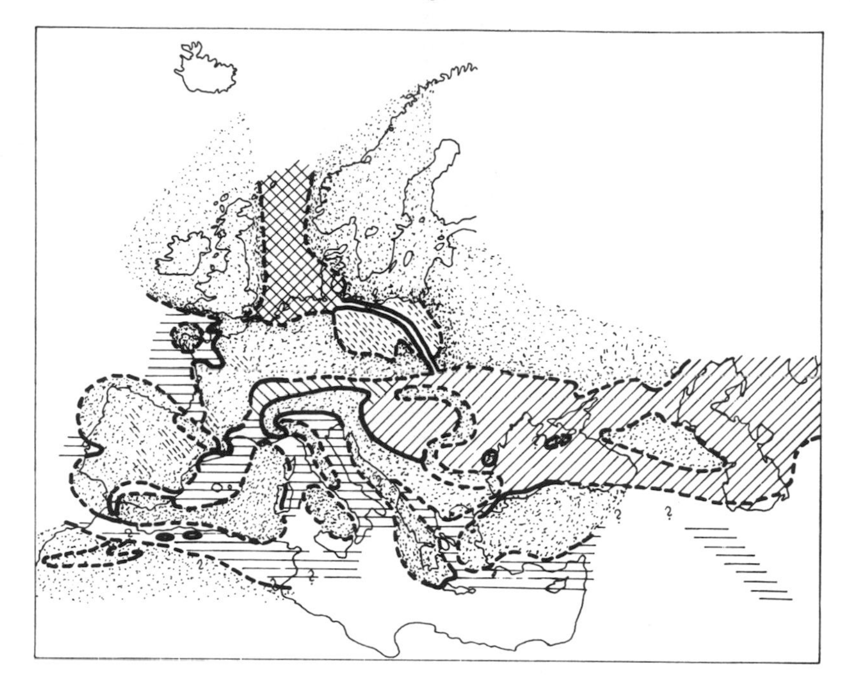

11. Paleogeographic map of Europe in mid-Miocene times, about 20 million years ago, shows a continent still partitioned by great interior seas and lakes.

dilians, and many strange and ungainly looking creatures inhabited the land. The time when man was to arise was still in the very distant future, and his ancestors were small, apelike creatures that were confined to the African continent.

In the limestone areas of Miocene Europe fissures and caves riddled the rocks, providing shelter for many mammals, birds, and reptiles. The fissure at Wintershof-West was a favored den for small carnivores, who left their bones and teeth in profusion in the earth that gradually filled the cavity. True, there are also the remains of some big bear-dogs—or dog-bears—that seem to have inhabited the cave from time to time. But the majority of the remains are those of small weasel- or skunklike animals, cats, and viverrids—early relatives of the mongooses and genets of today.

Among the small carnivores we also find remains of a creature about the size of a fox terrier, but which is neither dog, cat, or weasel. Latter-day scientists have given it the name *Ursavus elmensis.* It is with this small creature that the true bear line of evolution may be said to start. Actually, we can go still further back in time; we know the ancestor of *Ursavus,* but it is more doglike than bearlike, and so many scholars place it in the dog family. With *Ursavus* we come to the first animal definitely reckoned to be a bear, though indeed a very small and primitive one.

According to its remains, little *Ursavus* still had all its premolars, and they were slicing teeth of a truly carnivorous cast, just as in a dog. Its carnassials, on the other hand, were already taking on a bearlike look, and the molars show the beginning of the expansion of chewing surface that was to characterize the bear teeth of later times.

12. Life restoration of *Ursavus,* the earliest bear, by Margaret Lambert.

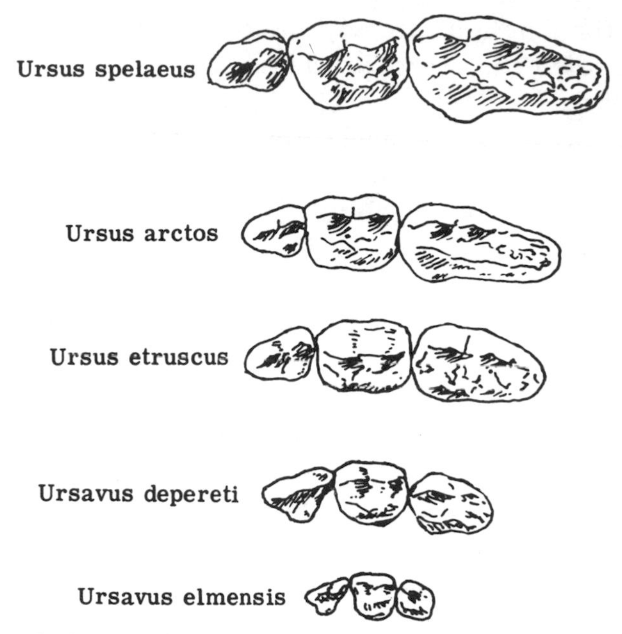

13. Upper cheek teeth of various bears, showing the progressive lengthening of the molars and the increase in size from the early *Ursavus elmensis*. *Ursavus depereti* was the last and largest of the *Ursavus* line. The Etruscan bear, *Ursus etruscus*, gave rise to the brown bear, *Ursus arctos*, and the cave bear, *Ursus spelaeus*. After Kurtén.

Unfortunately, we know very little of *Ursavus* except its teeth and jaws. Did it climb trees or live on the ground? There is no direct evidence, but we know that most bears are good climbers unless they grow too big and heavy. This trait is likely to be an inheritance from a common ancestor. We know, too, that the area in which *Ursavus* lived was probably forested. So it seems very likely that *Ursavus* did a bit of climbing. As for his diet, it probably featured insects and small vertebrates as well as vegetable matter; its feeding habits may have resembled those of the present-day badger.

The history of *Ursavus* continues through the Miocene epoch, which lasted many millions of years. As the aeons pass, we can see how the ursavi tend to become gradually larger and at the same time how their molars tend to expand and become more bearlike. But meanwhile, the landscape and its other inhabitants changed too. Mostly, of course, all of these changes were extremely slow. Thousands and thousands of years would pass, and still all is the same. It is only with the passing of a hundred thousand or a million years that changes may become barely perceptible.

But there are dramatic interludes. One of them came about 18 million years ago, still in the time of *Ursavus elmensis,* the dawn bear. At that time, a remarkable group of animals, which until then had been confined to the African continent, succeeded in making its way to Europe and Asia; later on we also find this group in North America. These were the mastodonts, early relatives and ancestors of today's elephants. From the middle Miocene on, these big creatures were the dominant animals of Eurasia, and they evolved into many different species.

The next remarkable invasion, occurring many millions of years later—12.5 million years before the present—came not from Africa but from North America. This continent was for millions of years the center of evolution of horses. But from time to time, a new and improved model of horse crossed the tenuous land connection between Alaska and eastern Siberia and made itself at home in the Old World. The late Miocene migrant was the famed *Hipparion,* which rapidly spread far and wide into Asia, Europe, and Africa and speedily crowded out those older editions of the horse theme that had entered the Old World earlier in the Miocene.

The hipparions were the last of the successful three-toed horses, with three hoofs on each foot instead of only one as in all living horses, zebras, and asses. They remained in vigorous possession of Eurasia for over 10 million years, until that time when the migration of newfangled one-toed horses from North America was let loose upon the Old World, and the hipparions went into a decline.

The early hipparion-fauna of the Old World coexisted with the last and largest of the primitive *Ursavus* bears. But at this time there was also living a somewhat more advanced type of bear. The only evidence of its existence discovered so far is a single tooth. It comes from Spain, from the remarkable site of Can Llobateres on the outskirts of the thriving city of Sabadell, near Barcelona. The immense harvest of fossil bones from this location includes at least seventy species of mammals, among which the abundance of gibbons and large apes is particularly striking. Clearly, when all these animals were living 12 million years ago, Can Llobateres was still a subtropical area with great forests inhabited by many remarkable beasts now long extinct.

The bear, as noted, is represented by a single molar tooth, but it happens to be one of the most distinctive in shape. The Can Llobateres bear was clearly more advanced than the contemporary ursavi, yet had not attained the evolutionary grade of a true *Ursus*. So we have called this creature *Protursus simpsoni*. It was an animal about the size of a sheepdog; it probably descended from one of the earlier *ursavi*, such as *Ursavus elmensis;* and it was probably ancestral to *Ursus*. Little more can be said about it, until scientists find more material.

About 10 million years ago there was a marked change in world climate, which started to become drier. In many areas the mighty forests that had existed for so many millions of years died out, and savannas, steppes, and deserts spread widely. During this process, many of those animals that were adapted for a life in the forests fared badly. But those that lived on the open plains—antelopes, hipparions, and many others—entered a golden age.

Bears are generally forest-living animals, and so we lose sight of some of them for the next few million years. We do know that the last and largest of the ursavi struggled on for some time before becoming extinct. We also know that the great *Indarctos* bears, which had arisen as an early side-branch from the ursavi, spread into North America. But the evolutionary line that was initiated by the *Protursus* of Can Llobateres escapes our search, and the Miocene epoch will have come to an end before we meet it again.

We are now in the Pliocene epoch of earth history, starting some 5 or 6 million years ago. Our stage moves to Roussillon and Perpignan in the south of France (and a contemporary site in Hungary), where we find the first member of the genus *Ursus*. He bears the name *Ursus minimus*, and he is indeed the smallest known member of his genus and the most primitive too. He probably reached about the size of the living sun bear or Malay bear, which is the smallest of the living bears.

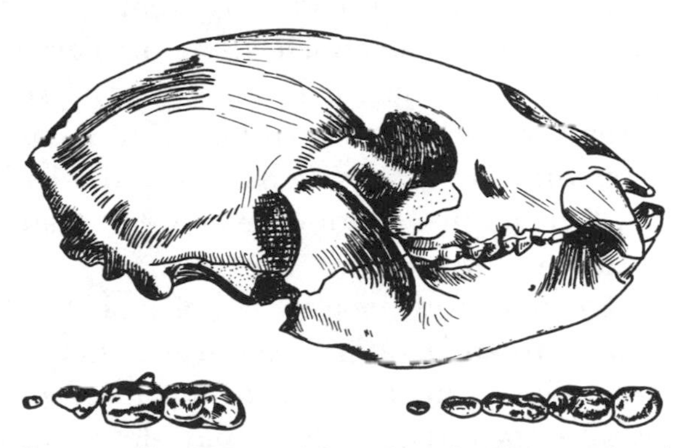

14. The small ancestral bear, *Ursus minimus*, from the Pliocene of southern France. The drawing shows the skull and jaw as preserved, and details of the upper (bottom left) and lower (bottom right) cheek teeth. After Viret.

Except for size, the resemblance between *Ursus minimus* and the Malay bear is not very close. For example, if you look closely at a Malay bear, you may note its remarkably stout eyeteeth. In contrast, the Pliocene *Ursus minimus* has slender, gracile canines. It also has a complete set of premolars, retaining their sharp, slicing character, although they are much less prominent than in the ancient ursavi. The grinders, on the other hand, have become more enlarged. So we can see how the trends initiated so many millions of years earlier are still going on, very slowly and gradually, towards the condition of the true bears.

At the time of *Ursus minimus,* the world was already on the threshold of the Ice Age. The climate was cooler than in the Miocene epoch and, as centuries and millennia passed, there was a slow swing between cooler and warmer conditions. In the high mountains and in the far north, ice caps waxed and waned with this secular climatic rhythm.

The forests where the first *Ursus* lived were quite different from the subtropical world of the Miocene ursavi. They were of the temperate type, containing deciduous trees and conifers. Palms were now unknown north of the Pyrenees and Alps, and crocodilians were gone from the streams. With the Pliocene epoch, the long Tertiary period finally came to an end. And far to the south, in Africa, small bands of a remarkable, new sort of biped were already moving about on the ground, using stones and sticks to hunt small game. But the first encounter between man and bear was still in the distant future.

As the Ice Age drew nearer, it was as if the tempo of world events were accelerating. About 4 million years ago, large oxlike animals made their first appearance on the scene: they later gave rise to bison, buffalo, and wild oxen. Somewhat later, a new breed of trunk-bearing animals migrated from Africa into Eurasia—the elephants and mammoths. A new chapter of earth history was indeed being written. We call it the Villafranchian age, the prelude to the Ice Age.

In the early Villafranchian age we still see *Ursus minimus* around, although he has changed a bit. He is somewhat larger than in the old times, and there have also been small, all but imperceptible, changes in his teeth. Bears of this kind were widely distributed in the Old World, and recent finds in North America show that this species, or a very closely related one, was present there too. The two living species of black bear, the American (*Ursus americanus*) and the Himalayan (*Ursus thibetanus*) are probably descended from *Ursus minimus.*

During the early Villafranchian age, about 3 million years ago, one-toed horses spread from America into the Old World, and rapidly penetrated Eurasia and Africa. The old hipparions lingered for some time but were gradually ousted, and eventually they vanished.

By mid-Villafranchian times, some 2.5 million years ago, changes had proceeded far enough for *Ursus minimus* to give rise to a new species: the Etruscan bear, *Ursus etruscus,* the typical bear of the later Villafranchian. This species, whose characters are known through numerous remains in Spain, France, and Italy, was also present in China. In the flesh it probably resembled living black bears, though of course its color is unknown to us. Through the Villafranchian age this trend towards large size continued in the Etruscan bear; late Villafranchian forms are larger than mid-Villafranchian ones.

15. Skull of the Etruscan bear, *Ursus etruscus,* from the Val d'Arno in northern Italy. Closely related to the living brown bears, it had a somewhat more primitive dentition. Original in the Natural History Museum, Basle.

The world of the Etruscan bear was already one in which continental icefields developed from time to time, only to melt again as the climate ameliorated. The Ice Age was nigh, and the pendulum swung between fully glacial conditions on one hand and interglacial conditions on the other, when the climate was as warm as now, or warmer. And it is in one of the early interglacials, about 1.5 million years ago, that we meet the last of the Etruscan bears.

This interglacial is called the Tiglian, after the old Roman name for Tegelen, a site in the Netherlands where rich deposits from this age have been found. The Etruscan bear is now as big as the living European brown bear, but it still carries the full complement of premolars (albeit very small ones) as an inheritance from the ancient ursavi and their doglike predecessors.

With the Tiglian interglacial, the Villafranchian age may be said to have come to an end: the prelude is over, and we are in the true Ice Age, the Pleistocene epoch of earth history.

The Tiglian comes to a close. In the Alps mountain glaciers grow larger, coalesce, and send icy tongues down along the valleys. They grow ever greater and finally engulf the mountain range, with only a few bare peaks protruding out of the frozen waste. Glaciation is upon the world again, and this glaciation is called the Donau (Danube).

Many thousands of years pass with great tracts of the earth's surface as if immobilized in icy stillness. Then, once more, comes the swing. Glaciers melt and retreat, and areas recently under ice emerge to be conquered by plants and animals.

Among those animals who return to the lands of their Tiglian ancestors we find the descendant of the Etruscan bear. Again, evolution has taken a step forward, for he has changed. The anterior premolars, already very small in the Etruscan bear, are now almost gone; in some individuals all are lacking, but many retain one or more of these vestigial structures.

Accompanying this change there is a tendency to a doming of the forehead, foreshadowing the cave bear condition. This new species of bear is called Savin's bear, *Ursus savini,* and it lived about 1 million years ago in the interglacial termed the Waalian. Its remains have been found in various sites, for example Bacton in East Anglia, England, and the Hundsheim fissure in Austria. Although large and impressive enough, these early cave bears were still much smaller, on average, than the true cave bear.

European and North American Glacial-Interglacial Sequences: A Tentative Correlation

Europe [1]	*North America*
Weichselian (Würm)	Wisconsinan
Eemian	Sangamonian
Saalian (Riss)	Late Illinoian?
Holsteinian	?
Elsterian (Mindel)	Early Illinoian?
Cromerian	Yarmouthian
(Günz)	Kansan
Waalian	Aftonian
(Donau)	Late Nebraskan
Tiglian	?
(Biber)	Early Nebraskan

[1] Names in parentheses refer to Alpine glacial terminology.

The pendulum swings again: cold conditions are back. The Günz glaciation reached its culmination some 800,000 to 900,000 years before the present. There is some evidence that in one of the cold phases that make up the Günz, longer-limbed bears, perhaps from the East, pressed into Europe to supplant temporarily the stocky-legged Savin's bear, but the problem of whether this intruder was a distinct species or just a steppe race of the early cave bear has not been settled. The latter alternative appears perhaps the most likely one.

The ice melts and the world is green once more as the wind of the Cromerian interglacial blows over the European scene. And man meets bear.

That encounter is only one of the remarkable things that happened in the Cromerian interglacial, which takes its name from the Cromer Forest Bed in East Anglia. It is a layer very rich in organic remains, including tree trunks, fossil beaver dams, and a copious number of bones.

From a chronological point of view, perhaps the most interesting event of the Cromerian interglacial is the reversal of the earth's magnetic field. Such reversals have happened at many times in the history of the earth, with intervals between lasting about a million years. Evidence of these reversals is found in the magnetic properties of the rocks that formed during a given interval. For instance, all of the rocks that form now carry the imprint of the "normal" polarity, while rocks formed in other times may have "reversed" polarity—north is south and south is north. The latest such reversal is known to have occurred 700,000 years ago in the age of the Cromerian.

There is evidence of Cromerian man in Europe at Mauer near Heidelberg, Germany, where a rich interglacial fauna quite similar to that of the Cromer Forest Bed has been discovered. During the 1960s, in a cave by Petralona in Greece, the same association of human fossils and Cromerian mammals was found. At both sites, bear remains are very common indeed. Mauer and Petralona contain the earliest finds of human fossils together with bear remains.

The bones of Cromerian man reveal a type of human still very primitive in many respects, yet closely related to us and probably our direct ancestor. He resembles the contemporary and better-known men of east and southeast Asia called *Homo erectus* (formerly *Pithecanthropus*), but also bears a certain resemblance to the late Pleistocene Neandertal men of Europe, to whom he presumably was ancestral. Of what happened between man and bear in the Cromerian we have no evidence.

The bear of the Cromerian may well be regarded as a

full-fledged cave bear. True, it is still a little smaller, although clearly larger and longer-jawed than its ancestor the Savin's bear. Also, the vaulting of its forehead is less prominent and its grinders are not quite as expanded as in the late Pleistocene animal. And so it has been given a species name of its own, and is known as Deninger's bear, *Ursus deningeri.* But there is much to say for just regarding it as an early, primitive race of *Ursus spelaeus.* We may compromise by calling it Deninger's cave bear.

But there is no rest for the pendulum of climate. Again there is a swing to cold conditions—the Elster glaciation—and then back to warm—the Holsteinian interglacial. We are now roughly 300,000 years before the present, and the bear in existence is *Ursus spelaeus* without any doubt. His remains have been found in caves in Germany and France, and especially interesting is the find of a good skull in the river gravels at Swanscombe outside London, England, which have also yielded a skull of early man.

The recounting of the long evolutionary history of the cave bear line may seem tedious, but it should give some understanding of how complete the evidence of its evolution really is. From the early *Ursus minimus* of 5 million years ago to the late Pleistocene cave bear, which became extinct only a few thousand years ago, there is a perfectly complete evolutionary sequence without any real gaps. The transition is slow and gradual throughout, and it is quite difficult to say where one species ends and the next begins. Where should we draw the boundary between *Ursus minimus* and *Ursus etruscus,* or between *Ursus savini* and *Ursus spelaeus?* The history of the cave bear becomes a demonstration of evolution, not as a hypothesis or theory but as a simple fact of record.

In this respect the cave bear's history is far from unique. Many other fossil sequences of the same type are known. Even in the case of such elusive fossils as our own ancestors, scientists are gradually obtaining evidence of their evolution, although much remains to be done.

I have treated the cave bear line as the main stream of bear evolution; this is perhaps justified, not only because the cave bear is the particular topic of this book, but also because it represents the culmination of the *Ursus* line—"the most bearish of bears." Still, it has been noted that the *Ursus* line threw off two side-branches, leading to the American and Himalayan black bears respectively. In some respects these bears remained near the evolutionary level of the Etruscan bear. We must now take note of another branch: the brown bear, *Ursus arctos*.

This animal differs from the Etruscan bear in its generally larger size, its somewhat longer and slenderer limb bones, and the greater reduction of its front premolars—to take the most important differences. The first brown bears in the fossil record appeared in China, in deposits dating back approximately to the Elster glaciation, about half a million years ago. Thus it would seem that the European population of Etruscan bears gave rise to the cave bears, whereas the Asiatic population evolved into brown bears.

Such a division of one ancestral species into two daughter species is quite a common occurrence in evolution. If the ancestral species is widely distributed, as was the case with the Etruscan bear, different parts of the far-flung population adapt to different circumstances and gradually tend to become more unlike each other. Thus develops a situation in which the species, so to speak, changes gradually from one end of its range to the other. There are even situations in which the end members of the chain look and act like distinct species in spite of being linked to each other by an unbroken line of interbreeding populations. There are many examples of this in present-day fauna. One of the most famous is that of the herring gull and the lesser black-backed gull, which in Europe look and behave like two quite distinct species. Yet they are joined by a chain of interbreeding populations through Siberia and North America, so that it is not really possible to say where one species ends and the other begins.

In such a situation, it may happen that a population, which forms a link in the chain, becomes extinct. The two branches of a species thus become isolated from each other and so become fully distinct. With the Etruscan bear, this situation probably occurred as a result of one of the early great glaciations. The ice sheets pushing south from Scandinavia and along the Ural Mountains may have isolated the European population from the Asiatic one, thus leaving them free to pursue separate evolutionary paths.

At a much later date the Asiatic daughter species, the brown bear, invaded Europe. From the Holsteinian interglacial on, or since about a quarter of a million years ago, both the cave bear and the brown bear were present in Europe. The brown bear also invaded North America, where it now lives in the guise of the grizzly and (Alaskan) brown bears. They are now regarded as local races of the species *Ursus arctos,* and their history in North America (except Alaska) is quite short, geologically speaking.

We still have to account for one species of the *Ursus* line. This is the polar bear, *Ursus maritimus,* which is now highly specialized for a completely carnivorous mode of life, but which in its anatomy carries a clear brown bear heritage. The polar bear may have arisen from brown bear populations on the Arctic coast of Siberia, which specialized in seal hunting. It apparently is the youngest of all the living species of bears, for the earliest known polar bear find is less than 100,000 years old (and incidentally comes from Kew in London).

Thrown into such (for a bear) completely novel circumstances, the polar bear probably evolved quite rapidly, under strong pressure by natural selection to adapt it to its new mode of life. This evolution is evidently still going on, for fossil finds during the last 10,000 years show that there have been distinct changes in some of its characters, although in an evolutionary perspective this is a very short span of time.

In the modern world there are three other species of bear, which do not belong to the *Ursus* line. The sun bear, or Malay bear, has already been mentioned. This small species, no larger than *Ursus minimus* of the Pliocene, is yet quite distinct in its anatomy, and so is regarded as a genus of its own, *Helarctos*. Its history is only known from the Ice Age when it already looked very much like the living form.

The second species is the sloth bear of India, "Baloo" of Rudyard Kipling's *Jungle Book*. It also belongs to a genus of its own, *Melursus*. A medium-sized bear, the sloth bear looks anatomically somewhat like an overgrown *Ursavus*, but again we have no reliable information on its history before the Ice Age.

The third living species of bear is the Andean bear of South America, *Tremarctos ornatus*. It is the sole remnant of a once-mighty host of American bears, whose history will be recounted in a later chapter.

It may help to visualize the history of the *Ursus* line if we see it in a diagram. In geology, successively younger strata are of course piled on top of each other. And so, in a geological diagram, the time scale is always vertical: the youngest are on top and the oldest at bottom. Another characteristic of such a diagram is that there is more detail near the top than further down: the closer we approach our own time, the better the evidence at hand. It is natural that we should be better informed about what happened 20,000 years ago than about events 20 million years ago.

Evolution of the Bears of the *Ursus* Line

Age (years)	*Ursavus Line*	*Cave Bear Line*	*Brown Bear Line*	*Polar Bear Line*	*Black Bear Lines*
Now (0)			U. arctos	U. maritimus	U. thibetanus U. americanus
End of Ice Age (10,000)		U. spelaeus			
Holsteinian (300,000)		U. spelaeus	U. arctos		
Cromerian (700,000)		U. deningeri			
Waalian (1,000,000)		U. savini			U. thibetanus
Tiglian (1,500,000)		U. etruscus			U. americanus
Mid-Villafranchian (2.5 million)		U. etruscus			
Early Villafranchian (3.5 million)		U. minimus			
Pliocene (5 million)		U. minimus			
Late Miocene (10 million)	Ursavus	Protursus			
Early Miocene (20 million)	Ursavus				
Pre-Miocene	Doglike Ancestor				

NOTES

Earth history in the Tertiary and Quaternary, with special reference to the mammals, is discussed in Kurtén (1971, 1972). Wintershof-West is treated in Dehm (1950); Can Llobateres in Crusafont and Kurtén (in press). The history of *Ursus* can be found in Erdbrink (1953); see also Kurtén (1968, 1969a). Cooke (1973) gives a radiometric chronology of the Pleistocene. For material on Swanscombe see Ovey (1964). The case of the herring gull and black-backed gull is discussed in Huxley (1942); the evolution of the polar bear in Kurtén (1964).

CHAPTER FOUR

The World of the Cave Bear

The Ice Age started well over a million years ago, and it is still going on.

Precisely when it started is difficult to say; it did not come on overnight. There was a long transition, a gradual cooling of the climate that can be traced back for many millions of years. In the end this cooling led to the formation of great continental ice sheets, like those still covering Greenland and Antarctica. But there was also a swing between cold and warm, between glacial and interglacial phases. We now live in one of the latter phases—one of the warm intervals within the Ice Age—but it will be followed by another glacial phase, perhaps in 10,000 years or less, unless we succeed in cooking ourselves before that by unleashing too much nuclear energy.

In the previous chapter we followed the history of the cave bear line up to the appearance of the species *Ursus spelaeus,* some 300,000 years ago. This was the time of the Holsteinian interglacial. Since then there have been two

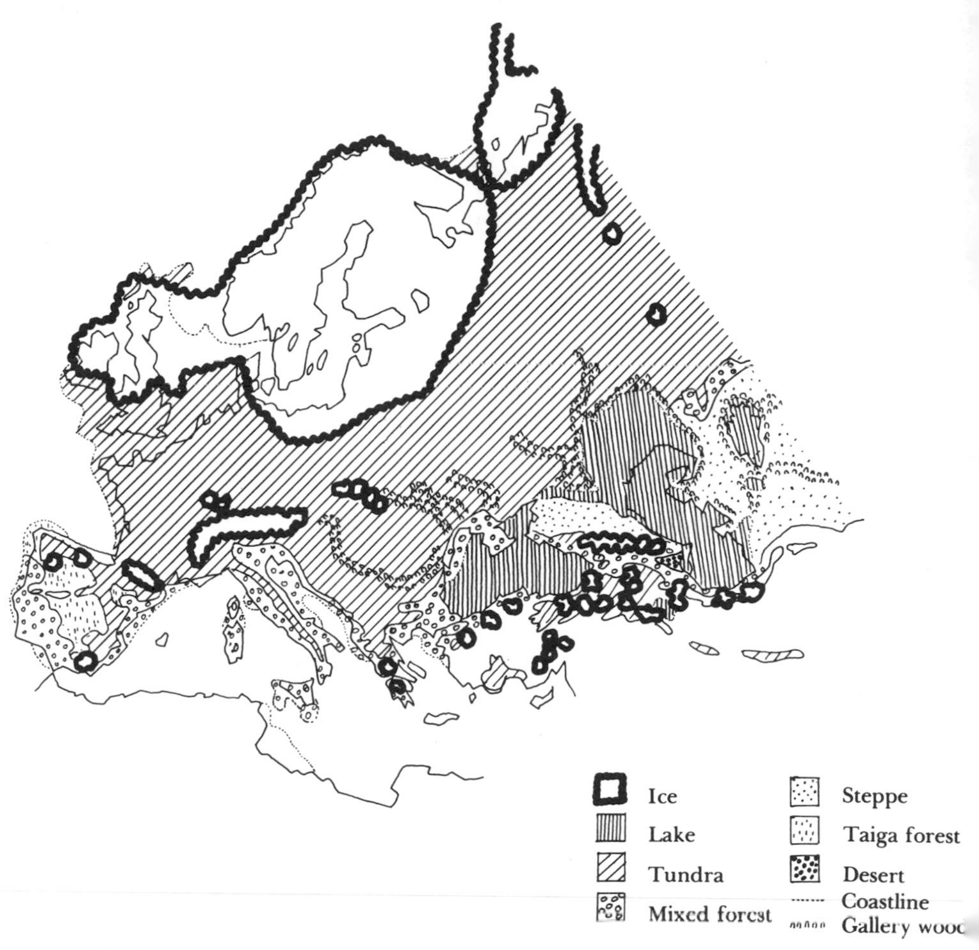

16. Geography of Europe in Weichselian times shows the northern parts of the continent covered by great icefields; smaller glaciations have formed on the mountain ranges further south. A large area is tundra; the southern peninsulas are partly forested. The sea has receded, exposing the bottom of the North Sea and the English Channel; the Caspian is much enlarged, flooding the plains of southern Russia. Data from Königsson.

complete swings in climate. The first was the Saalian glaciation, during which the land ice of Europe covered a greater area than ever before or since. It was followed by a very warm interglacial, the Eemian, which occurred about 80,000 years ago. Some 70,000 years before the present, the last glaciation, the Weichselian, got under way and lasted up to about 10,000 years ago. The period from the Holsteinian interglacial to the end of the Weichselian glaciation was the time of the true cave bear in Europe.

Glacial, interglacial, glacial: there is a grand swing of the pendulum. But, in a way, such a description is an oversimplification. There were many shorter-range oscillations too: the interglacials are not uniformly warm nor are glacials uniformly cold. The climatic history of the Weichselian glaciation has now been mapped in great detail, and we find

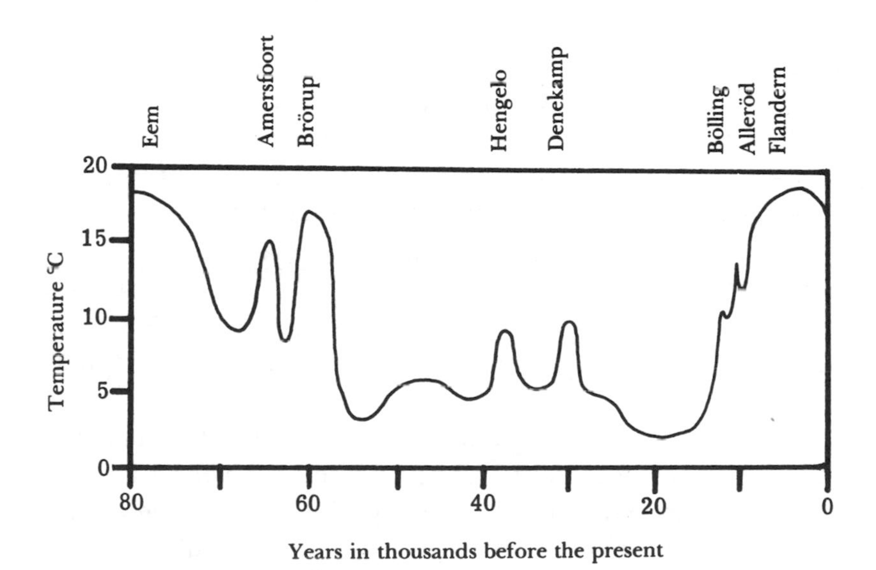

17. Mean July temperatures for western Europe during the last 80,000 years of earth history. Names of warm oscillations are indicated, beginning with the Eemian interglacial and ending with the Flandrian interglacial, which is still going on. Intervening warm phases are termed interstadials. Data from van der Hammen.

an incessant shifting between colder and warmer phases. Particularly notable are the several warmer spells of short duration, termed interstadials, which interrupt the glaciation. Although the last 70,000 years or so of earth history are dominated by the last glaciation, there have been warmer interstadials at various times. These begin with the early interstadials called Amersfoort and Brörup, which occurred only shortly after the first onset of the glaciation, and are somewhat uncertainly dated to 60,000 to 70,000 years ago. Next came the Hengelo and Denekamp interstadials between 40,000 and 30,000 years ago; during this time the Neandertal man vanished, and men of modern type appeared in Europe. After this phase came a very cold stage of the glaciation indeed, when once more central and southern Europe were overrun by mammoths and other arctic forms, and the reindeer was the favorite prey of the hunting tribes. Finally, the climate definitely started to improve; there were two premature warmer intervals, the Bölling and the Alleröd, both followed by setbacks with increasing cold, but beginning 10,000 years ago present interglacial conditions have prevailed in earnest.

Studies in America and Asia have revealed a very similar story, and the dating of events indicate that probably the climatic shifts were synchronous all over the Northern Hemisphere, and apparently in the Southern Hemisphere too.

It was, then, a far from stable world that the men and animals of the late Pleistocene inhabited. Viewing that span of time from our foreshortened perspective, it seems that there was a turmoil of change, a continuous shifting of climatic zones, a waxing and waning of glaciers. To individual beings of the time, of course, such changes would mostly be too slow to be perceptible. Yet there may have been times when glaciers advanced or retreated fast enough to be noticed by those creatures of long memory that lived close by, just as in our day glaciers in Canada, Alaska, and Scandinavia have been noted to waste away in response to a short-term warming of climate—one of the minor oscillations, too slight to be reflected in the 80,000-year curve.

So we should not imagine the cave bear in just one kind of setting. Rather, there was a shift between two extremes, the warm interglacial and the cold glacial. Perhaps I might first introduce the reader to the warm world, that of the Eemian interglacial, some 80,000 years ago. In some ways it will seem less strange than the glacial world that is closer to us in time (see figures 18–19).

The climate was warm, probably even warmer than today, for at the maximum of the Eemian, many southern plants and animals ranged far north of their present-day limits. A particularly spectacular case is that of the hippopotamus, which, following the great Danube, Rhone, and Rhine rivers, reached the Thames and finally got as far

18. Restoration scene of Europe in the Holsteinian interglacial, about a quarter of a million years ago, shows Merck's rhinoceros, straight-tusked elephants, and giant deer, in open park landscape. The same species survived in the Eemian interglacial, 80,000 years ago; but the giant deer of that time belonged to a different subspecies with antlers spread more horizontally, not deflected in a curve as in this Holsteinian form. Restoration by Margaret Lambert.

19. British river scene in the Eemian interglacial, with cave hyenas, hippopotami, bison and (in the distance) straight-tusked elephant. The skull of fallow deer may be seen at the entrance of the cave, left foreground. Restoration by Margaret Lambert; after Sutcliffe.

north in Great Britain as Stockton-on-Tees at 54½ degrees north. But there are numerous other cases that prove that the temperatures were high in Eemian times. The sea level was higher than it is today, encroaching on what is now land, and this condition probably also contributed to the equability of the Eemian climate.

The record of interglacial life has been unearthed from caves, river terraces, lake-bottom sediments, ancient bogs, and the like, wherever the remains of plants or animals are likely to be preserved. Apart from a few exotic species, the plant life of the Eemian would not look too unfamiliar, except for its exuberance, which would put most present-day European forests to shame. Eemian animal life, however, was quite another thing.

Besides the cave bear there were many other large mammals, many of which are now extinct. The largest creature in the land fauna of the Eemian was the straight-tusked elephant. The name alludes to one of its characteristics: the almost straight, very big tusks, which emerged far apart from a very broad rostrum and diverged widely. The stature of this elephant was immense, reaching as much as 15.7 feet (4.8 meters) at the withers. The European form is usually called *Elephas antiquus,* but it may be just a race of the Asiatic *Elephas namadicus.* These fabulous elephants are now long extinct, but the living elephants of Africa and, especially, India are not too distantly related to them, and probably give some idea of what they looked like.

The elephant and the cave bear certainly met each other in Eemian Europe; we find their fossil remains together in various places, such as the interglacial spring deposits in the Weimar area of Thuringia, Germany.

The other great pachyderm of the Eemian was the rhinoceros. There were actually two species, of which Merck's rhino, *Dicerorhinus kirchbergensis,* inhabited the woods and park forests that were also the homeland of the straight-tusker. Merck's rhino was related to the living Sumatra rhino, *Dicerorhinus sumatrensis,* and like that species carried two horns in tandem fashion on its nose. In open country was a related species, *Dicerorhinus hemitoechus,* which appears to have been a grazer, whereas Merck's rhino browsed off trees and bushes.

Among the many other large herbivores in the fauna was the now-extinct bison or wisent, *Bison priscus,* also found in Asia and North America. Larger than any living bison, it also differed from them in its appearance, which we know from accurate paintings by Ice-Age man. This bison had a peculiar double-humped appearance, perhaps due to the development of a dark mane on the back of its neck in front of the shoulder hump.

Besides the bison, although less common, was the aurochs or wild ox, which is extinct as such but survives in the domestic cattle of today. The last wild aurochs was killed in 1627.

The deer tribe was well represented in the Eemian by the red deer (which was a member of the same species as the wapiti or American elk), the moose (or true elk), the fallow deer, and the roe deer. All were large forms, rivaling or exceeding in size the largest present-day races of their species. Still larger was the now-extinct giant deer, sometimes called Irish elk but actually related to the red deer or wapiti. Its Latin name, well-founded indeed, is *Megaceros giganteus,* or "gigantic great-antler." The enormous palmate antlers reached maximum spans of up to 11.5 feet (3.5 meters) and may have been the bulkiest and heaviest head ornaments ever sported by a member of the deer family. These impressive structures were, in fact, used to impress: they were display organs that were shown to their greatest advantage when the creature stood with its head raised, facing its enemy or rival.

Wild boar and wild horses were also found in the forests of the Eemian, and the main river systems, as we have noted, were inhabited by hippopotamuses surpassing those of the present day in size.

With so much big game around, there were naturally large predators as well. The greatest of them was the lion, which in the late Pleistocene was the most wide-ranging wild mammal of all time. Apart from its present-day homelands in Africa, it also lived in Europe, all through northern Asia, North America, and even in South America as far south as Peru. Never before or after has a mammalian species, except man and his followers, attained such a worldwide distribution. The northern lions were great creatures, markedly larger than any present-day lions. They are known as cave lions, because their remains are so often found in caves.

There was also a cave leopard in the Eemian fauna, but its remains are less common. Again, it was a large form of the species. Smaller members of the cat family alive in the

Eemian included the lynx and the wildcat. Very scarce finds indicate the existence of a great scimitar-toothed cat, *Homotherium latidens,* through the Eemian and into the Weichselian glaciation.

The most common carnivore in the fossil record of this time, except for the cave bear itself, was the big cave hyena. This form is now regarded as a large race of the living African spotted hyena, *Crocuta crocuta,* which in Pleistocene times ranged widely in Africa and Eurasia; besides Europe its remains have been found in India, Tibet, China, and Korea. Hyenas, formerly regarded as cowardly carrion-eaters and nothing else, are in fact fiercely aggressive hunters, usually operating in packs. They must have been a common sight in the Europe of the Ice Age. The hyena often made its lair in caves, where the bones of these predators may occur by the thousand, together with those of their prey.

The remains of wolves are common in Eemian deposits, and indicate that Eemian wolves resembled those of the present day. Occasionally, bones of the dhole or red hunting dog are also found in Pleistocene beds; this species is now extinct in Europe.

The cave bear's relative the brown bear is also common in the Eemian fossil record. In the spring deposits or travertines near Weimar (Ehringsdorf), which have yielded cave and brown bear remains, there is a record of still another species of bear: the Tibetan black bear, *Ursus thibetanus,* today found only in Asia. As we have noted, this species of bear diverged from the cave bear line at the *Ursus minimus* stage.

Of course, there were also many smaller mammals in the Eemian animal world. Beavers and polecats inhabited the streams, squirrels and weasels the trees; red foxes and badgers dug their dens in the ground. Most of the animals were typical of forest or parkland. Both the fauna and flora of the Eemian give a luxuriant impression: it was a world of plenty, and its inhabitants grew large and sleek.

Living off this plenty we find Neandertal man as the lord of creation in Eemian Europe. His bones, too, are found in the travertines by Weimar—as well as at numerous other sites—and many of the animal bones of these beds are spoils of Neandertal man's hunt.

With an arsenal featuring early Mousterian points and hand axes, and fire-hardened yew spears, Neandertal hunters followed the game. Rather short of stature, they were powerfully built with remarkably large heads, whose beetling brows, protruding faces, swept-back foreheads, and bull necks distinguish them from modern men. However, we should not picture Neandertal man as an apelike monster, the way he is shown in some life restorations. He was a human being without any doubt, and many fossil finds reveal him engaged in truly human behavior—his most endearing activity was perhaps his custom of laying flowers on the graves of his dead. This was revealed recently by Dr. Ralph Solecki, who discovered the custom while digging a Neandertal cave in Iraq.

All things pass. The Eemian comes to an end. Where a thousand years ago grew the magnificent mixed oak forest, pine and birch are now the dominant trees, and as other millennia roll by, they too are gone to the South, and the tundra is back. The world is in the grip of the Weichselian glaciation. Immense volumes of water are bound up in land ice, and so the level of the ocean drops by 300 to 400 feet; the climate becomes not only colder but also drier and more continental.

There are many oscillations. Occasionally the coniferous forests advance far north at the heels of waning ice sheets; but the ice and the tundra come back. Many of the warmth-adapted interglacial species are gone, others have found refuge in the Levant and in Africa. But many of them also withstand the cold, and stubbornly continue their existence in Weichselian Europe. The cave bear is one of these.

As the straight-tusker might be seen as a symbol of the interglacial fauna, so the woolly mammoth embodies the glacial. In stature he was rather less impressive than the giant elephants of the interglacial. The Weichselian mammoths rarely exceeded 10 feet (3.3 meters) over the withers, and many living elephants are bigger than that. What he may have lacked in size, he more than compensated for in his remarkable appearance, which was a favorite motif of many cave artists, and which we can also reconstruct from partly frozen carcasses discovered in the permafrost areas of Siberia and Alaska. The humped shoulders, sloping back, peaked head, and gigantic curved tusks make an unforgettable ensemble of characteristics.

By the Weichselian glaciation the forest rhino, Merck's rhino, was gone and in its place we find the great woolly rhinoceros with its long, shaggy fur. Musk oxen, now surviving only in the American arctic, were a common sight on

20. A chance encounter between mammoths and wolves in Weichsel-glacial Europe probably ended in the rapid dispersal of the wolf pack. The only serious enemies of the mammoth were man and the rare scimitar-toothed cat. Restoration by Margaret Lambert.

the Eurasian tundra. In the deer family, reindeer was predominant on the tundra but in wooded areas moose and red deer were common. Mammoths, reindeer, and woolly rhinos, and perhaps also the *Bison priscus* probably made seasonal migrations between their two principal environments: the tundra or the loess steppe in summer and the taiga (open coniferous forest) in winter. Canadian reindeer still make such migrations, and the scale is spectacular. In this way, such arctic animals may have become associated with mainly boreal-temperate creatures like the moose and the aurochs.

Most large carnivores are hardy, wide-ranging animals well equipped to withstand the rigors of a cold climate. A present-day example is the tiger in Siberia. We tend to think of lions and hyenas as denizens of the tropics, but in fact they thrived in Weichselian Europe. If they responded to the cold, they did so by growing even larger. Thus the hyena tribe culminated in the great true cave hyena, *Crocuta crocuta spelaea,* of which thousands of bones have been unearthed in caves such as Kent's Cavern, dug so long ago by

21. In Europe during the Weichselian glaciation, lions and hyenas may have disputed the right to a carcass just as in Africa of the present day; but the prey would have been a different animal, such as the young woolly rhinoceros shown in this life restoration. Restoration by Margaret Lambert.

Pengelly, and the Teufelslücken or Devil's Hole near Eggenburg, Austria, explored by the great caver and naturalist Krahuletz.

Regular dens of the cave lion are less common but there is at least one example in the cave Wierzchowska Górna in southern Poland. There, remains of numerous lions, ranging in age from cubs to old adults, show that the cavern was actually inhabited.

The cave bear and the brown bear, the wolf, hunting dog, and red fox, all inhabitants of the Eemian forests, stayed on in the changed setting of the Weichselian; but we now also find such colder climate predators as the wolverine and arctic fox. And many of the small mammals were such that are now only found in the far North: the lemmings, the varying hare, and the redbacked voles. From the East, hardy steppe animals spread with the great loess steppes into central and western Europe: the bobak marmot, the jerboa, the steppe lemming, and the steppe pika.

The reign of Neandertal man in Europe continued well into Weichselian times. It came to an end about 35,000 to 40,000 years ago, around Hengelo and Denekamp interstadial time. Henceforth, we find men of modern type in Europe—so-called Cro-Magnon men, people of our own species, *Homo sapiens*. Whether they evolved from the local Neandertalmen or migrated into Europe from somewhere else, crowding out Neandertal man in the process, is still disputed, and the problem need not concern us here.

Like their predecessors, these new peoples still followed the hunting trade. But their weapons were soon improved; an important innovation was the atlatl or spear thrower, which added greatly to the range and force of the spear (the bow and arrow were not yet invented). Culture became richer and even more artistic than that of Neandertal man whose handiwork reveals an appreciation of beauty. With Cro-Magnon man a still stronger striving for beauty and perfection came to the fore. This is the time of the masterpieces of cave art. To a student of the animal life of the Ice Age, cave art is especially significant, for it presents eyewitness recordings of the wildlife of the time—including many species that are gone forever and which we can never see in the flesh.

So life continued to flourish under the cold skies of the last glaciation. Of course, some of the Eemian species disappeared, though many of them, like Neandertal man, managed to hang on during the earlier half of the Weichselian. This is true for the straight-tusker and Merck's rhino, and even the hippopotamus held out, in Italy, into the early Weichselian.

In fact the losses were surprisingly small, considering what a tremendous change in climate occurred with the transition from interglacial to glacial conditions. Presumably what had really happened was that the Pleistocene species had already survived many such transitions and had evolved adaptations to withstand them. Those unable to do so had died out much earlier. A very resilient, adaptable fauna was being fashioned by natural selection. The cave bear was one of those animals who stayed on, unabashed, in their chosen area, glaciation or no glaciation.

For a large carnivore like the cave bear, this area was surprisingly small. The range of the species *Ursus spelaeus,* and of its predecessors *Ursus deningeri* and *Ursus savini,* was almost completely confined within the bounds of the European continent (of which Great Britain formed a part during glacial epochs, when the English channel was dry land). The northern boundary of the range extends from southern England to the East, barely taking in the southernmost parts of the Netherlands, then through Germany and southern Poland into the southern parts of the Soviet Union. Fossils of the species are still common in the Caucasus and recent discoveries also indicate that it may be

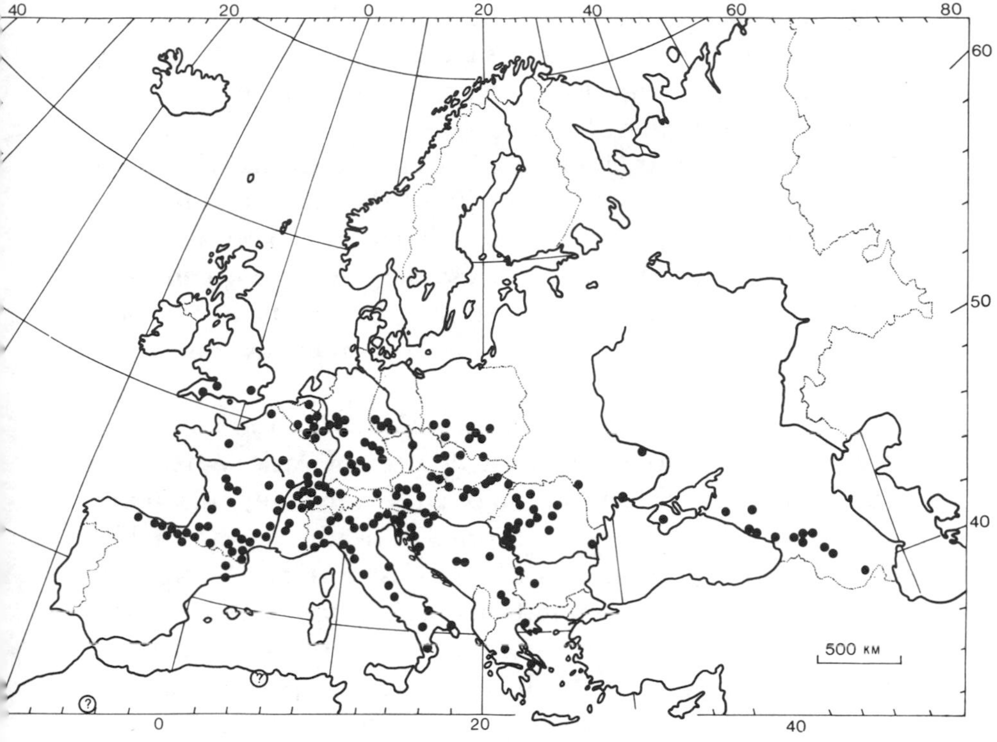

22. Distribution of the cave bear, *Ursus spelaeus,* in the Pleistocene. Each dot represents one or more sites with fossils of the species; two North African records are uncertain. Only Holsteinian and later records are included. The actual number of sites is many times greater than the dots, and some sites may contain the remains of hundreds or thousands of bears.

found north and east of the Caspian Sea, in the southern Urals, and in Kazakhstan. In the South, it ranged into northern Spain, almost all of Italy, and Attica in Greece.

It is true that from time to time, cave bear remains have been reported in areas outside this range—for instance in various British and Irish caves and even as far afield as in China. In the cases I have been able to investigate, these have been mistaken identifications: the animal in question was in fact a large brown bear. An exception may be formed by a couple of finds in Morocco and

Algeria, which certainly look very much like the cave bear; so it is possible that the species temporarily ranged to North Africa. But the majority of bear fossils from this area are certainly brown bear, *Ursus arctos*.

The cave bear, then, was an almost exclusively European species. Another peculiarity of the species is that it is almost exclusively found in caves. There are exceptions, but they are few and there are no instances of large numbers of cave bear bones occurring in river or lake beds; in contrast, a large proportion of the *Ursus deningeri* remains come from such deposits.

Caves of the sort that we find cave bears in were formed in limestone and related types of rock. Where the subsoil water is rich in carbon dioxide, the limey rock tends to dissolve and cavities are formed beneath the water table, especially along existing cracks and fissures, and along contacts between different kinds of rock. In some areas, veritable rivers flow for miles along such subterranean channels, perhaps to emerge in the form of a great spring. With the sinking of the water table, intricate cave systems of enormous proportions may be laid dry, and men or beasts may settle in them. Other types of caves are formed at the seashore by the pounding of the waves against limestone cliffs; such caves are smaller, but when the sea level goes down and the cave is high and dry, it may also be used as a living site.

Was the cave bear very dependent on caves? And in what way? These questions have been disputed, but, although fossils of the species are also found in open-air deposits, it is notable that, from about the Holsteinian on, the cave bears get to be extremely scarce at such localities. Moreover, the true cave bear does not occur at all in areas where caves are uncommon or absent. Either the species was dependent on the caves as such, or else it was dependent on the hilly or mountainous, often forested type of environment in which caves occur. Probably both relationships were in fact true. The nature of the dependence will be discussed in later chapters.

How about the areas outside the species range as we know it today? Can we expect that future discoveries are going to extend the range substantially? This is certainly a possibility in the eastern part, from Lake Aral to the east, where paleontological prospecting is still in its early stage. It seems to be much more unlikely for the well-studied Pleistocene faunas in Europe, Africa, and the Middle East, and also in Siberia and China. There are, for instance, plenty of fossiliferous caves outside the cave bear range in England, Wales, Ireland, Germany, Poland, Spain, Italy and so on, but there are no cave bears in them, while brown bears may be common. At least in this area, I think we have a pretty definite idea of the range limits of *Ursus spelaeus*.

NOTES

Standard texts on the Pleistocene include Charlesworth (1957), Flint (1971), Woldstedt (1969), and Zeuner (1959). Pleistocene mammals of Europe are discussed in Kurtén (1968) and Toepfer (1963). Kurtén (1972) is an illustrated history of the Ice Age. Modern restoration of *Bison priscus* is discussed in Geist (1971). The travertine faunas near Weimar will be treated in a series of publications issued by the Institute of Quaternary Paleontology of that city. Solecki (1971) calls Neandertal men "the first flower people." The material on cave bear ranges is from Koby and Schaefer (1961), Wolff (1938–1941), and original; material concerning eastern extension on preliminary data can be found in Kozhamkulova (1974). Arambourg (1933) found possible cave bears in North Africa.

CHAPTER FIVE

Males and Females, Dwarfs and Giants

No two bears are alike, any more than two human beings.

In his remarkable study of seventy-six cave bear skulls from the Dragon Cave at Mixnitz in Styria, Austria, Dr. W. Marinelli distinguished among large skulls and small skulls, domed skulls and flat skulls, elongate, narrow, greyhoundlike skulls and short, broad, puglike skulls, various combinations of these types, and practically every stage in between the extremes. He concluded that all of this bewildering variety was still within the bounds of a single species, thus confirming the opinion of Georges Cuvier expressed more than a century earlier (see figure 23).

Marinelli's study revealed another remarkable situation. The small bear skulls were concentrated in the upper strata of the Mixnitz cave. This suggested to Marinelli and his fellow workers that the cave bear had evolved into a dwarf race just prior to its extinction.

At the same time, Bachofen-Echt, working with the iso-

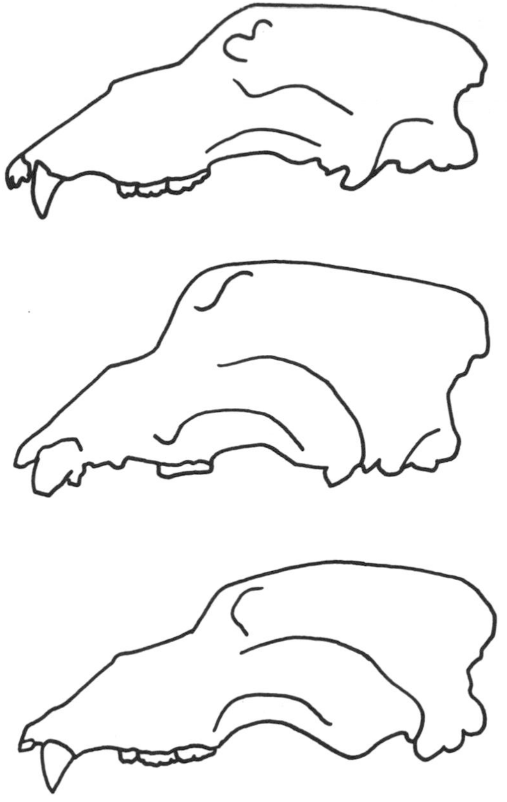

23. Individual cave bear skulls show striking variations in shape, especially in the profile of the forehead. At one time, they were thought to belong to different races or even species, but in fact the whole range of variation may be found in a single local population.

lated canine teeth found in such profusion in the strata of the Mixnitz cave, reported that there were two distinct size groups. He suggested that the large teeth were male and the smaller ones female. We might expect the number of males and females in the population to be equal, and in fact Bachofen-Echt stated that this was so in the lower cave strata. In the upper strata, however, male canines tended to be more numerous, finally outnumbering the female canines three to one!

Marshaling these and other data, Professor Othenio

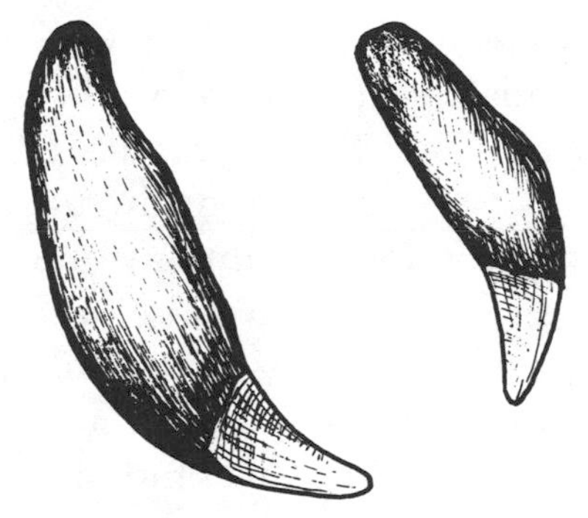

24. Upper canines of male and female cave bears, both drawn to the same scale to show sexual dimorphism in size. Most canine teeth of the cave bear are easily identified as to sex.

Abel, the leader of the Mixnitz excavations, suggested a startling but logical explanation of the various phenomena, which could also be developed into a theory for the extinction of the cave bear. The keyword in Abel's explanation is domestication; not, however, domestication by man, but rather a sort of self-domestication resulting from the sheltered life and lack of enemies of the cave bear. In such circumstances, natural selection would be radically reduced, and all sorts of inadaptive variation might occur. The great variation in size and shape, the dwarfing, the increasing preponderance of males, all suggesting that something had gone wrong with the reproduction and heredity of the species—this evidence was seen by Abel as symptomatic of a serious degeneration. The high frequency of various diseases, some of them perhaps due to the unwholesome microclimate of the cave, was also seen as a degenerative symptom. In the end, Abel postulated, the degeneration became so extreme that the species was unable to survive any longer.

Various aspects of this theory (with which I disagree) will be discussed in the chapters to follow. At the moment, the variation seen in the cave bear will be my topic.

What is variation in an animal species? The species

exists in a multidimensional world of space and time, and consists of individuals united by bonds of heredity and descent, forming an incredibly complicated bundle of intertwining strands. We can now distinguish among no less than five types of variation, within any one species and they should be kept separate, or we will get into a muddle.

The first type of variation is that within a single individual, growing from a fertilised egg to an adult animal. To evade the blur introduced by this kind of variation, we have to limit our comparisons to animals in the same stage of life—for instance, fully grown ones.

Second, we have individual variation. No two individuals (except for identical twins) have exactly the same genes; no two live all their lives in precisely the same environment. In 1804 Rosenmüller advanced the factor of individual variation as a partial explanation of the great variability of the cave bear.

Third, there is sexual variation. Males and females of a species differ, mostly not just in the reproductive organs but also in various other characters. In many mammals, the males tend to be larger than the females, and this certainly holds true for present-day bears. Rosenmüller, again, was well aware of this fact and pointed to it as an additional cause of variation in the cave bear.

The fourth and fifth kinds of variation are due to evolution: variation in space and variation in time.

Variation in space is seen, for instance, when you compare the rather small brown bears of northwestern Europe with the very large ones in eastern Asia. They have become different by evolution, but they are still connected with each other by a series of interbreeding populations. They form what Sir Julian Huxley termed a cline, or a character gradient, in space: as you go from West to East, the size tends to increase.

Tip the gradient over into the time dimension, and you

get evolutionary change in a single area. For instance, in western Europe the size of the brown bear was rapidly reduced in the millennia just after the Ice Age. This is a cline in time, and it illustrates our fifth type of variation. The dwarfing of the cave bear at Mixnitz, if true, would be a case in point.

Let us begin with the question of male and female. Is it really possible to tell, just from a fossil bone, the sex of its possessor?

Bachofen-Echt had noted a difference in the size of the canine teeth (see figure 24), but unfortunately his criteria for determining sex were rather loose and were not backed up by a detailed analysis. A few years later, Dr. Karl Rode made a series of measurements of bear canines, both in the cave bear and the brown bear, and was able to show that both species showed a distribution into a large and a small type of canine tooth, with only a few intermediates. He interpreted them as male and female, respectively.

It remained for Dr. Koby, in 1949, to clinch the matter by studying canines of modern brown bears, in which the sex was known, and comparing size in the two sexes. Measuring the width of the canine at the base of the crown, he found that the average width for male bears was 2 millimeters greater than that of female bears. He also took the same measurement in a sample of 682 fossil cave bear canines and found that they formed two distinct frequency curves with peaks about 6 millimeters apart. Thus, in the cave bear, the male would have a canine about 6 millimeters broader, on average, than the female. Such a difference, in other organs than those of reproduction, is termed a secondary sex dimorphism.

We must now ask ourselves how these results are influenced by the types of variation other than sexual variation, which I have listed. As far as the canines are concerned, the effects of individual growth can be ruled out, for a fully formed tooth crown does not change in size. The only change possible occurs when the tooth is worn by use, and it is of course quite easy to spot this condition and avoid measuring such teeth.

Variation due to evolutionary change can be avoided if the sample is drawn from a limited area and period of time. This is fairly easy to do in the case of modern species. Koby's cave bear sample is somewhat less homogeneous. It came from a limited area in the Jura Mountains, and so is spatially very restricted, but the time element is a little uncertain. The material probably dates from the greater part of the last glaciation and may represent a time range of as much as 50,000 years.

Still, the curves Koby obtained are very similar to corresponding curves for local populations of living bears—say, the polar bears of Greenland or the big brown bears of Kodiak Island. It really seems that the cave bears of the Jura Mountains were relatively stable in size during the last glaciation.

In later bear studies, the homogeneity has been increased by taking samples for separate sites representing only shorter time intervals. In all these cases the curves were of the same type—bimodal, with good separation of males and females. In such frequency curves, then, the separation of the two peaks represents the sexual dimorphism, while the scatter around each peak represents the individual variation. A curve for the Mixnitz cave bear turns out to be quite similar (see figure 25).

Is the size difference limited to the canine teeth? They have been intensively studied because a large amount of isolated teeth can be got together. But wherever we find teeth still sticking in the jaw or skull, we find that large canines are in large skulls and small canines in small skulls. The sexual dimorphism in size is, in fact, seen in all of the body. Almost any skeletal dimension will show a bimodal

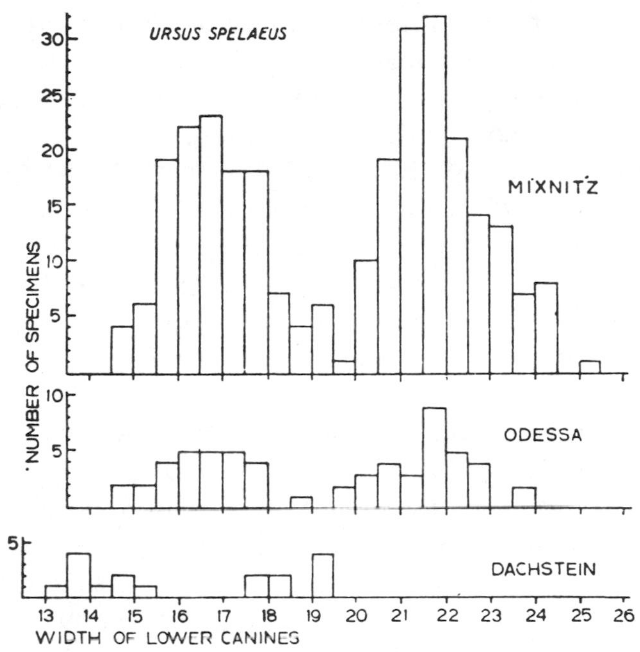

25. Width of lower canines in samples of cave bear shows distribution into two size groups, representing male and female individuals respectively. In the large Weichselian bears from Mixnitz and Odessa, male canines average about 22 millimeters in width (see scale at bottom), female less than 17 millimeters. In the smaller Eemian cave bears from the Dachstein cave, male canines average 18–19 millimeters in width, female canines about 14 millimeters. After Kurtén.

frequency curve of the same type as the canines (the cheek teeth are an exception). In effect this means that, as soon as we have established the typical male and female dimensions for a given bear population, practically every bone may be determined as to sex.

Such a size difference between the sexes is common in some groups of carnivores—the bears, the cats, and also the weasel family; on the other hand, it is weak or nonexistent in dogs and hyenas. In the primates, males are often larger than females; this trait is especially marked in the larger monkeys and apes, but is to some extent true for man as well. In the even-toed ungulates—such as buffalo, oxen, antelope, and deer—there is the same kind of difference. Cases in which the female is larger than the male are also known, although they are less common in mammals (in some cetacean species the female is larger than the male).

Study of the Mixnitz "dwarfs" revealed that they are not dwarfs at all but normal females. In every instance, they turned out to have canines that fitted into the female size group (the sole exception being a youngish male that would have grown to normal male size had it lived). So one of the supposed degeneration traits of the cave bear proved to be illusory.

The same is true for many other supposed dwarf races of the cave bear. Indeed, dwarf "subspecies" of the cave bear have been formally named on the basis of quite normal female specimens. One case, involving two caves called the Sibyllenhöhle (Sibyl's Cave) and the Hohlestein (Hollow Stone), both situated in Württemberg, Germany, is instructive and will be recounted here. Dr. Karl Dietrich Adam, of the State Natural History Museum in Stuttgart, told me the story of these collections.

The cave of Hohlestein was excavated in the 1860s by the museum in collaboration with the Swabian Cave Society (Schwabischer Höhlenverein). A great number of cave bear fossils were found in this typical bear cave, and they were then distributed between the museum on one hand and the society on the other. The selection, however, was made by the museum authorities, who naturally kept the finest specimens for the museum. In practice, this meant the large and showy ones, or, in other words, the males. The less impressive specimens, or females, went to the society and into various private collections, and have since been lost to science. The museum collection now at hand from Hohlestein consists of 90 percent male skulls and jaws, and only 10 percent female! Needless to say, this does not represent the original relationship. Some time ago, a Hohlestein specimen was acquired by the museum from one of the few old private collections still in existence: as might be expected, it was the jaw of a female bear.

Several years later, in 1898, the Sibyl's Cave was excavated, once more in collaboration between the State Natural History Museum and the Swabian Cave Society, and the material was again distributed between them. But this time the society had the pick, and the result was exactly what might be anticipated. The museum was left in possession of a good collection of female cave bears, while the fossils of the male bears went into the society members' private collections. In this case, too, most of the distributed specimens have been irretrievably lost to science. The museum collection from the Sibyl's Cave comprises 23 percent male and 77 percent female specimens.

These quite normal females form the basis for the description and naming of a "dwarf race" of cave bear, called *Ursus spelaeus sibyllinus.* There is, of course, no such race.

Such fantastic disproportions between the numbers of males and females, then, arose from bias in collection. How about other caves?

Working through the material from Mixnitz preserved in the collections in Vienna, I was unable to confirm the extreme disproportion of three males to one female, stated by Bachofen-Echt. The actual numbers in this collection—including skulls, jaws, and isolated canine teeth—were found to be 346 male specimens and 230 female. The relationship of males to females is about 60 to 40, with a moderate preponderance of males.

Now it should be remembered that this material only constitutes a very small fraction of all the specimens that actually were present in the cave. The digging of the Dragon Cave was not primarily a scientific but a commercial undertaking: the cave sediments were mined for phosphate. In all such cases, scientists are in a very difficult position, trying to save as much valuable material as possible, yet being forced by circumstances to select and to discard.

Now, any kind of selection will result in a statistically biased collection—as we have clearly seen in the cases of Hohlestein and the Sibyl's Cave. Since only a small part of the material can be collected, and much of the collecting may be done by untrained persons, it is evident that any kind of striking, showy, or unusual specimens will be more readily taken than less impressive ones, which nevertheless are necessary for the population background.

In the case of the cave bear, whole skulls and jaws, big bones, large isolated teeth, fossils showing unusual pathological changes, and so on, tend to become overrepresented in collections, while smaller isolated teeth, fragmentary specimens, and the like, are underrepresented. It was estimated that the entire fossil material in the Dragon Cave at Mixnitz comprised the remains of some 30,000 to 50,000 bears; only a small fraction of these remains could be collected.

Mining richly fossiliferous sites for phosphate must now be regarded as predatory exploitation, for the number of such sites is limited and each phosphate-mining operation results inevitably in large-scale destruction of scientific documentation.

Apart from such exploitation, the simple fact that cave bear material is plentiful at many sites has often fostered an unscientific attitude. Selection has been made of suitable museum specimens, "perfect" skulls, unusually large bones, or other things that caught the investigator's fancy, while other material was refused, left to destruction, or distributed to the public. Any refusal of specimens and any selection, on any grounds whatever, automatically results in a biased sample and distorts or destroys the scientific value of a collection. Even when all the material is saved, field notes may be available only for the best-preserved specimens.

Fortunately, authoritative and competent work is now being done by many scholars, who are fully·aware of the unique interest and possibilities of cave fossils for population studies, in numerous countries. In this the local speleological societies have a great responsibility, which I think they clearly recognize.

We return from this digression to the question of male and female cave bears at Mixnitz. Now, in the degeneration theory of Abel it was suggested that a change in the heredity of the species led to a superabundance of male births relative to female.

The inheritance of sex has been well understood for a long time. We know that in mammals, including man, sex is determined by the so-called X and Y chromosomes—or sex chromosomes—in the nuclei of the egg and sperm cells. In a female mammal, the two sex chromosomes are of the same type (X), and the egg cells that are formed in her reproductive tract will contain one X chromosome each. The sex of the offspring is determined by the sperm cell, for the male, having one X and one Y chromosome in the nuclei of his cells, will distribute them equally to the sperm, one-half of which will carry X chromosomes, while the other half carries Y. The fertilization of an egg cell by a sperm carrying an X chromosome will result in the female XX combination; if the fertilization is made by a sperm carrying a Y chromosome, the result is a male XY.

A disproportion between the sexes could result from an unequal production of X and Y sperm. This situation is unlikely, because the actual production of sperm occurs in such a way that equal numbers are necessarily formed. Again, the X-carrying sperm could be less viable than the Y ones, thus not reaching the egg. Actually, the opposite seems to be true, for X sperm tend to live longer than Y sperm. On the other hand, Y sperm are a little lighter and move a little faster, so they have a slightly greater chance of reaching the egg. But all these mechanisms operate in mammals other than the cave bear, without causing a markedly differential birthrate.

In the living grizzly bear of Yellowstone Park, for instance, a slight excess of male cubs is produced, but this is apparently offset by slightly higher mortality, so that, among adult grizzlies, females are somewhat more numerous than males (54 to 46 percent).

Was there, then, actually a differential birthrate at Mixnitz? The sample of skulls, jaws and canine teeth from Mixnitz—a total of 576 specimens—represents individuals of all ages. Some 195 belong to young individuals, from one-year-old cubs to half-grown, not yet sexually mature bears. It is in this group that the differential birthrate ought to be reflected. However, out of the young bears, 103 could be sexed as male and 92 as female.

This evidence does not bear out the theory of a strongly differential birthrate. The slight excess of males among the cubs is not really significant in the statistical sense; that is, it could be accidental, just as we might get heads 103 times and tails 92 times when we flip a coin 195 times. Even if the difference is real, it might just mean that the mortality of male cubs was slightly higher than that of female cubs, a situation quite common in mammals.

The analysis of the Mixnitz collection shows, then, that the excess of males occurs almost entirely among adult bears. In this group, the males outnumber the females almost two to one; the actual numerical relationship is 243 to 138, or 64 and 36 percent.

What about other caves? Koby observed that females were somewhat more frequent than males in his material taken from the Jura Mountains. In his largest sample, consisting of 456 canine teeth from the cave of Gondenans-les-Moulins, male fossils comprised some 44 percent of the total. Lower male percentages were found at other sites, such as Vaucluse (36 percent), Montolivot (33 percent) and Saint-Brais (only 28 percent). On the other hand, there are many caves where the ratio between male and female fossils is close to the "normal" 50-50. Among material tested, this holds true for the caves of Odessa and Nerubaj in southern Russia, the Austrian Dachstein cave, the Swiss cave of Cotencher, and the Spanish cave, Cueva del Toll.

I think that this situation can be explained, not as the result of varying birthrates, but as due to the active choice of caves by the bears themselves. We must not imagine that the bear just stumbled into any cave that happened to be handy and went to sleep. On the contrary, the bears must have been connoisseurs of caves and known exactly what they wanted. Furthermore, it seems likely that a bear, once it had selected a cave, would tend to return to the same site from year to year and to defend it against intruders. This behavior is called territoriality, and we have some reason to suspect it in the cave bear, as I hope to show later on.

It is hardly a coincidence that the cave of Saint-Brais, in which females outnumber males almost three to one, is also quite a small cave, while the Mixnitz Dragon Cave, with a preponderance of adult males, is a huge one, and Cueva del Toll, where males and females occur in equal numbers, is a medium-sized cave. It is only natural that a female bear that is going to have cubs seeks a shelter that is small and easily surveyed and defended and tends to avoid large caves that serve as winter quarters for several individuals. Adult male bears can be quite dangerous to small cubs, even if their own offspring, and an experienced female bear would try to deny them access to her winter lair.

But let us return to the problem of the "dwarf" cave bear. Is it just a myth? Evidently not. At some sites, we do meet with distinctly smaller forms of the species.

That these are real "dwarfs," or members of a smaller race, is shown by the fact that both the males and the females are smaller than the males and females, respectively, of the "normal" form. A good example is the bear from the Schreiberwand Cave of the Dachstein Mountains in Austria. There is, for example, a sample of seventeen lower canine teeth from this cave, and they are clearly distributed into two size groups, which must represent males and females, respectively. However, the males of the Dachstein form are smaller than the males of the "normal" cave bear, and the females are smaller than the "normal" females (see figure 25). A comparison of other body dimensions produces similar results.

The Schreiberwand Cave is in the high mountains, at an elevation of about 7,200 feet (2,200 meters), and thus in an area that was covered by ice during the last glaciation. It could only have been inhabited in interglacial times. Small forms of the cave bear have also been found in other caves at a high elevation; indeed this situation is so common that Professor Kurt Ehrenberg has dubbed this type of bear the *"hochalpine Kleinform,"* or High Alpine small form. The bears from the Salzofen Cave by Bad Aussee, and from the Schottloch in Gosau, also in Austria, are almost as small as the Schreiberwand Cave bears; and the altitudes of these caves are only slightly less, or about 6,500 feet (2,000 meters). Like the Dachstein cave, these caves could only have been accessible in a warm climate, so the bears whose remains are found in them lived in interglacial times.

The "normal" type of cave bear, on the other hand, is not found at altitudes over 3,300 feet (1,000) meters) or so (which is the approximate level of the famous Dragon Cave at Mixnitz). Indeed most of the caves with such remains are found at altitudes of about 1,600 feet (500 meters) or less.

All of the "normal" or large cave bears evidently date from the last glaciation. In the lowlands are also found bear caves of earlier, interglacial date, and the bears from these tend to distinctly smaller dimensions. A good example is the bear from the Repolust Cave, studied some years ago by Dr. Maria Mottl of Graz, Austria. The Repolust cave bear has various primitive characters, such as the occasional presence of anterior premolars, a relatively low glabella, and so on, all of which bring the ancient Deninger's cave bear to mind. The same is true for the bear fossils from the basal levels of the Mixnitz cave, which are older than the typical or "normal" cave bears of the main cave earth stratum.

So we have a suggestion that the cave bear attained its full stature only with the onset of the last glaciation. Although much remains to be done in the way of careful measurement and painstaking analysis, it seems that this suggestion, originally based on Austrian material, is borne out by studies of fossils in other areas. In Britain, for instance, cave bear remains have been found only in a few caves in southern England (in Devon and the Mendips) and these mostly date from the last glaciation. These bears were, if anything, even larger than the "normal" form of central Europe. But there are also a few finds of older date. For example, the famous Kent's Cavern in Torquay contains a long sequence of strata, extending well back into middle Pleistocene times. The earliest deposits here have yielded a "dwarf" type of bear not unlike the High Alpine small form of Austria. The same is true for cave bear remains found in the ancient Thames terrace by Swanscombe, southeast of London. They date from the Holsteinian interglacial, about a quarter of a million years ago, and the same may well be true for the earliest deposits in Kent's Cavern. An analogous case in the USSR is illustrated in figure 26.

The matter now seems definitively clinched by the discovery of a richly fossiliferous bear cave, La Romieu in France, with a mid-Pleistocene sequence covering at least the Elster (Mindel) glaciation, the Holsteinian interglacial, and the Saale (Riss) glaciation. As revealed by Dr. François Prat of Bordeaux, there is here an evolutionary sequence

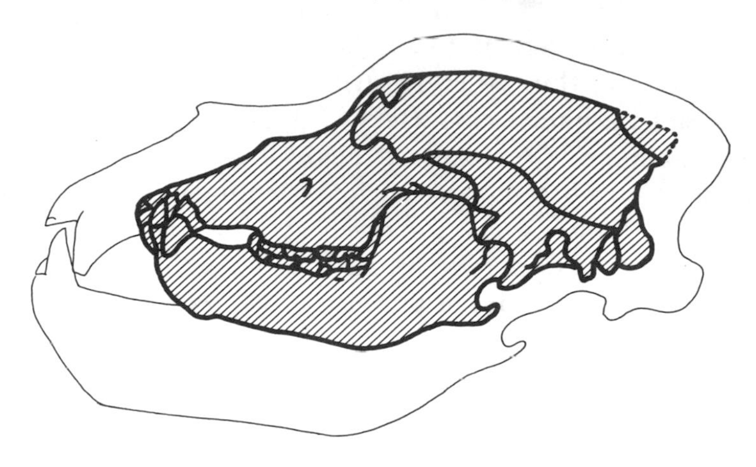

26. Skull of a small, probably Holsteinian-age cave bear from Krasnodar, USSR, compared with that of a large Weichselian cave bear (light outline). After Borissiak.

from *Ursus deningeri* in the earliest strata to relatively small forms of *Ursus spelaeus* in the Holsteinian and Saalian strata.

If we try to chart the size changes in the cave bear line, we find an early culmination in the Elster glaciation (about 500,000 to 600,000 years ago), with very large forms of Deninger's cave bear. This was followed by a size reduction in the Holsteinian interglacial, a slight increase during the Saale glaciation, and a renewed dwarfing in the Eemian interglacial. Finally, during the last glaciation, the species reached its maximum size, in the guise of the so-called "normal" form.

Oddly, a somewhat similar pattern of size change—an early size culmination in the Elster glaciation, followed by decrease and a second culmination in the last glaciation—may be seen in other carnivore species, such as the brown bear, the cave hyena, and the wolverine.

There is probably some common factor behind this regularity. In 1848, the zoologist Carl Bergmann pointed out that an increase in size will reduce heat loss through the surface of the body, and so would be adaptive in a cold climate. The cave bear in fact seems to obey Bergmann's Rule, at least partially: glacial forms average larger than interglacial. But the trend is somewhat irregular—there is no particularly marked size increase during the Saale glaciation—and so this is probably not the only factor involved. (Actually, there are also mammals in nature which vary inversely to Bergmann's Rule!)

As we have seen, then, there is a lot of temporal variation in cave bear history. But how about the spatial variation? Was there much variation between contemporary bears in different parts of the species range? Let us imagine ourselves transported back in time, say 25,000 years. Would there be notable differences between, say, British and central European cave bears? Or between those north and south of the Alps? Or even between those in two neighboring valleys?

The problem is a difficult one, for it is not always easy to pin down exactly the interval within the last glaciation in which the population of a certain site lived. What we can do is to compare, say, the dimensions of animals living at different sites. In some cases we do find differences in the mean values which, slight as they may be, are still statistically valid and point to some degree of racial differentiation—just as in other animals, including man. For instance, the cave bears which we find in southern Britain during the last glaciation tend to particularly large size. They appear to be veritable giants of their species. The largest lower jaw of a cave bear I have ever seen (though fragmentary) is a specimen from Kent's Cavern (and, the joke being on the natural history museum in London, this specimen is in the American Museum of Natural History in New York City). The cave bears of nearby Belgium are almost as large.

Gradients in space, or clines as Huxley named them, are also found in living bears, as we have noted. In areas where the gradient is particularly steep or, in other words, the change particularly rapid, we must assume that genetic interchange is relatively low. In the species *Ursus arctos,* for instance, such a steep gradient is found between the coast of Alaska and British Columbia on one hand, with its giant brown bears, and the inland grizzly population. The two do hybridize in nature, but the zone in which the hybrids occur is narrow and they never seem to become particularly common. In Europe, there is also a comparatively steep gradient between the bears of Scandinavia and Finland on one hand, and the bears in the central European USSR on the other.

In the cave bear, size gradients seem on the whole to be relatively steep, or, in other words, local populations tend to differ markedly from one another. This suggests a restricted gene flow between local populations. And so we may perhaps conclude that the cave bears, as individuals, were somewhat less wide ranging than brown or grizzly bears generally. As I have already mentioned, this trait could have been connected with territoriality—each bear would establish itself as the "owner" of a limited home territory, perhaps with the cave as its center. If true, this would be a marked difference from the related species *Ursus arctos,* which is not really territorial.

NOTES

For Marinelli's study of the cave bear skulls and also the papers by Bachofen-Echt, see the Mixnitz monograph (Abel and Kyrle, 1931). The extinction theory was set forth by Abel (1929). Sexual variation in bears is discussed in Koby (1949) and Kurtén (1955); the latter also treats local and temporal variation in a preliminary way. Sex distribution in various caves is discussed in Kurtén (1958). Mottl (1964) published on the bears from the Repolust and other Eemian sites in Austria. A study by Dr. F. Prat on the bears from La Romieu is pending. On sex distribution in Yellowstone grizzly bear see Craighead et al. (1974).

CHAPTER SIX

Man and Bear

What happens when man meets bear? Half a century ago, an amazing answer came from the Swiss Alps.

During the years 1917 to 1921 Emil Bächler, of the museum in St. Gallen, Switzerland, dug the Drachenloch Cave—one of the "Dragon's Lairs"—near Vättis in the Tamina Valley. The cave, at an altitude of 7,335 feet (2,240 meters) above sea level, forms a deep tunnel running more than 200 feet (70 meters) into the cliff. The deposit in the cave turned out to contain an immense number of cave bear remains, including several well-preserved skulls and complete limb bones. At that elevation, the site would have been inaccessible during the glaciation; thus the bears must date from the interglacial, the time of early Neandertal man in Europe.

To his surprise, Bächler came to realize that the skulls and bones were by no means scattered haphazardly. On the contrary, they seemed to be oriented rigidly in certain preferred directions. Could they have been deliberately placed by man? Soon there were further discoveries that made Bächler sure.

The finds in the Drachenloch were reported by Bächler in 1923 and in another report that he published seventeen years later. The most remarkable find was that of a large stone coffin or chest, containing a group of cave bear skulls and covered by a large stone slab. All of the skulls were pointing the same way. The coffin was about three feet (1 meter) high; the sides consisted of limestone slabs, which, like the cover, had originally fallen down from the ceiling of the cave. Unfortunately, in the course of the excavation, workmen destroyed the chests, and no photographs were taken.

It is even more unfortunate that Bächler's two sketches, published in 1923 and 1940 and purporting to show the chests and their situation within the cave, are quite contradictory. (They are reproduced in figures 27 and 28.) They agree in showing the chest resting upon layer V in the

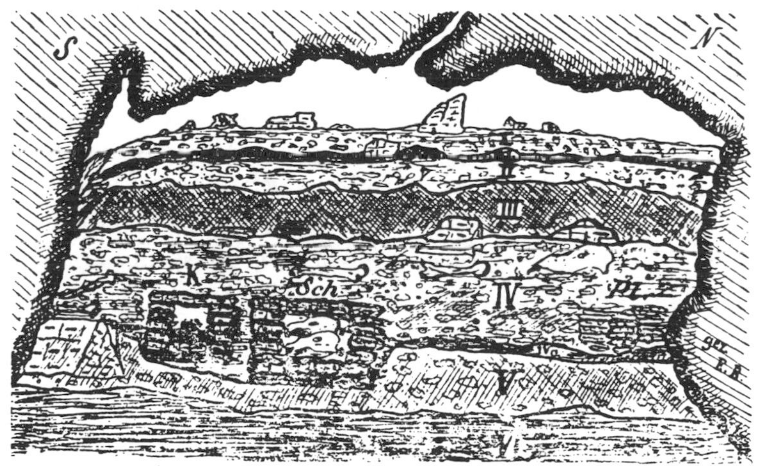

27. Cross section of the Drachenloch Cave as published by Bächler in 1923, showing stratigraphy of the deposits in the cave and the position of stone chests containing skulls and bones. For a discussion of this and the following figure, see the text.

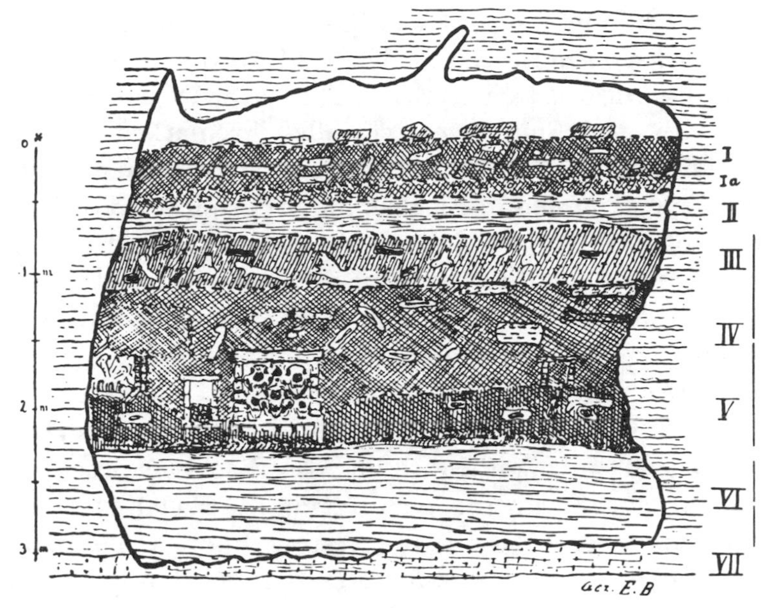

28. Cross section of Drachenloch Cave as given by Bächler in 1940. Compare it with preceding figure and see text.

sequence of strata in the cave (the layers were numbered from top to bottom). They also agree in the outline of the cave walls and ceiling, showing that both pictures are supposed to represent a north-south cross section. Otherwise, however, hardly anything is the same.

In the figure of 1923, for instance, a large chest is shown with two skulls seen in profile, facing south. Layer IV, which rests on top of the chest, contains various long bones and a skull facing the same way. Beside it is another, slightly smaller chest containing long bones. The walls of the chests are made of small, horizontal, even slabs resembling bricks.

In the 1940 picture, the big stone chest is still there, but this time it contains six or seven skulls, all of which face east (presumably a more orthodox direction, comments Koby). The second chest with the long bones is now much smaller, and, in addition, a sort of wall has appeared at the southern end, enclosing more bones. The walls of the chest are now built of vertical slabs, and the skull in layer IV has vanished. "Le myth est définitivement cristallisé," remarked Koby.

These are not, of course, the only finds at Drachenloch Cave that suggested to Bächler the ordering hand of Paleolithic man. There was, for instance, a bear skull with the thigh bone of another bear stuck through its cheek in such a way that it could only have been got in by twisting it around. In 1940, Bächler noted the find of a skull resting upon two parallel shin bones (tibiae). It is the same skull, but in 1921 the thighbone was stuck through the left cheek, and in 1940, through the right. Also, Bächler's description does not agree with the picture, for the sketch shows the whole arrangement of bones resting upon a flat slab of rock, while the description states that "curiously enough, this small deposit of bones did not have a stone slab for a base." Actually, Bächler was not present when the find was made.

Is there, then, any other evidence of the presence of man, apart from this curious arranging of the bones? In fact they are precious few. There are no flint implements. There are no burnt bones. There are no butchering cuts on the bones. All there is are some hearths, indicating that an occasional visitor or group made a brief stop. Any prolonged stay would certainly have been reflected in the sprinkling of numerous flints.

But how could such elaborate structures as the suggested stone coffins have come about if not erected by man? And the alignment of the bones? There is no question of a hoax. Bächler was known as an honorable, ardently patriotic man, and he certainly believed in the existence of the stone chests that he described.

To understand this situation we must go into how bear skulls and bones are actually preserved in caves. And the story begins with a hibernating cave bear dying in the cave (just *why* it died does not concern us for the moment; we shall return to that question in chapter 7).

It would sometimes happen that after death the great bear cadaver remained unnoticed by scavengers such as hyenas, wolves, and gluttons and was left to moulder away. As the soft parts disintegrated, great amounts of phosphate were produced. Now, the deposit in a bear cave is often very rich in this substance, which may make up as much as 50 to 55 percent of the total, and often is mined commercially. Bat guano, which is found in some bear caves, also contains phosphate, but the content is much lower, less than 10 percent. So the main part of the phosphate found in bear caves came from rotting bear flesh.

Skin and flesh gone, the cadaver winds up as a skeleton lying on the cave floor where the bear died. But this is only the beginning of its story; more about it presently.

It could also happen that scavengers did come across the body; they would eat the soft parts and pull the skeleton to pieces. Hyenas might smash some of the bones. Hyenas are known to swallow quite large pieces of bone, which are regurgitated after some time, more or less affected by bowel juices and movements. The result may be curiously suggestive of human interference: perfectly round holes may appear in the bones, pieces of bone may become wedged together as if intentionally, and so on. Also, the hyena-bitten bones splinter into sharp edges and so may take on the appearance of implements fashioned by man.

The end result of the scavengers' work is now a disarticulated skeleton scattered over the floor of the cave, the individual bones in varying states of disrepair.

In time the cave will get a new inhabitant, most likely another cave bear, which will enter it in the autumn to prepare for hibernation. The bones and fragments on the floor will be in the bear's way and will be trampled to pieces.

The larger objects, for instance such skulls and long bones that have not been broken into fragments, will be pushed to the side. Typically, they will finish up somewhere by the walls; as every bear cave explorer knows, most of the well-preserved skulls are found by the walls of the cave. The chance of surviving is particularly good if they get pushed into a niche that protects them from rockfall and other damage.

"In the Petershöhle in Germany a rock niche, situated like a cupboard in the rock, contained five Cave Bear skulls, two thigh bones, and one brachial bone," stated Professor Abel in 1935. He went on to say, "All these pieces must have been put into this niche by Ice Age Man, as a deposit formed by water is quite out of the question."

Of course it does not have to be a niche in the wall. Any kind of protecting rock will do. Such protecting niches may be produced at any place in the cave by rockfall from the ceiling. Percolating and freezing water gradually widens the cracks in the limestone that forms the bedrock of the cave. The cracks often form in the bedding plane of the limestone. In time, pieces of rock, some of them flattish slabs, are dislodged and fall down on the floor.

If there is already rubble on the floor, the rock may be left in a more or less vertical position and will be likely to protect the bones that get pushed in beside it. Further rockfall from the roof may occur in the same place, and in many cases will result in slabs being left in standing or semierect positions if they hit obstructions already present on the floor.

Meanwhile, the cave deposit is slowly built up by the dust brought in by animals, by the guano dropped by bats, which often roost in great numbers in caves, and by the products of the disintegration of the various animals that die in the cave. In time, the cave earth will also fill the interstices in the niches or "chests," and we arrive at a final situation that, with some moderate stretching of the imagination, may well be ascribed to deliberate burials.

It is evident that repeated pushing of such elongate objects as skulls, jaws, and long bones into niches or along walls will inevitably tend to align them in the same direction, suggesting that they were positioned by intent. In fact, all of the pushing, trampling, gnawing, biting, swallowing and regurgitating, pounding by falling rocks, and so on, which the bones undergo in a well-frequented cave—and which Koby comprises under the single term "dry transport" ("charriage à sec")—is likely to produce, from time to time, the most peculiar results. And we must remember that such freaks or oddities are precisely the ones that tend to be selected for survival by natural agencies. For instance, skulls in niches are likely to be preserved, while skulls in the middle of the cave floor will be trampled to fragments and survive only as isolated teeth and pieces of bone pushed down into the earth. It is estimated that some 30,000 to 50,000 bears died in the Dragon Cave near Mixnitz, but only some 76 good skulls were found. One skull out of 500 or thereabouts! No wonder skulls in bear caves *look as if* somebody had put them in a safe place.

Taking this possibility into account, it now seems impossible to accept the evidence for deliberate burials of bear skulls and other bones in the Drachenloch Cave near Vättis. The same goes for other sites, such as the Petershöhle in southern Germany, the Dragon Cave near Mixnitz in Austria, and the Wildenmannisloch in Switzerland, where deliberate positioning of skulls and bones has been claimed though not in actual "chests."

In the Petershöhle, for example, a great accumulation of skulls was found together with a lot of rocks; one skull was close to a hearth, but the bone showed no trace of burning. Of course the rock rubble would tend to protect those skulls that came to rest among the stones, so that the whole arrangement may perfectly well be due to natural causes. In the Mixnitz cave there is a lateral passage called the Abel Gallery, which was found to contain no less than forty-two bear skulls and many long bones. Here, too, man was supposed to have intervened, but Joseph Schadler considered natural causes sufficient to account for the accumulation.

In the Wildenmannisloch, Bächler found bear skulls with slabs of limestone resting on top, "making the impression of having been intentionally placed in a horizontal position," an impression that is rather weakened by the reflection that most flat slabs of rock will naturally come to rest in a horizontal position.

Enthusiasm for the "bear cult" is naturally contagious. Secondhand and thirdhand quotations from the original works often tend to glorious embellishment—to be found even in the writings of such a sober prehistorian as the Abbé Breuil himself, who once referred to the Petershöhle as a Paleolithic "tabernacle." In Abel's description of the Drachenloch there were "several stone chests," each containing four to five bear skulls, and there also were "numerous stone and bone artifacts" together with the bear remains, although in fact no flints at all were found. All this, according to Abel, proves that "during the Mousterian period in Central Europe, the killing of bears was accompanied by skull and long bone sacrifices."

The most trivial occurrences have been cited as evidence for the cult of the cave bear. It is known, for instance, that certain Siberian tribes reverence the bear and, among other things, extract certain teeth from the skulls. Thus the find of a cave bear skull without incisor and canine teeth may well bring to a mind sufficiently prepared by ardent belief the conviction that this is a Neandertal parallel to the modern practice. If you come with such an argument to a museum curator, the best you can hope for is a sad smile;

he will have profound knowledge of the ease with which certain teeth tend to drop out of drying skulls.

The bear feast of the Lapps, often mentioned in discussions of bear cults, was a hunting rite, not a sacrifice. After the ceremonial eating of the bear, the bones were buried, generally with at least the skull and some other bones in approximately correct position. Many such bear graves have been found, but they show no resemblance to the Alpine bear caves.

I believe that we must conclude, with Koby, that there is no real evidence for a cave bear cult among the Neandertal men who inhabited Europe in the last interglaical and the earlier part of the last glaciation. There *may* have been a bear cult—but we have no proof. It is the more to be regretted that the Drachenloch structures, whatever they were, were not properly documented, by means of photography, detailed plans, and so on.

When we come to the time of modern man in Europe—from about 35,000 B.P. to the end of the Ice Age some 10,000 B.P.—the evidence is somewhat better. And yet it does not tell us as much as we might hope, or as much as some students have claimed.

The art of Paleolithic man is often thought to have religious significance. It is generally agreed that this art dates from the latter half of the last glaciation, when Europe was inhabited by men of modern type—Cro-Magnon man and his successors. We find the remains of such peoples associated with a sequence of Late Paleolithic cultures, many of which excel in the arts of painting, engraving, and sculpture. Most of the pictures represent animals, and the majority are game animals of the type that would have been important in the economy of those hunting tribes—the bison, the wild ox, the ibex, the red deer, the reindeer, the horse, the mammoth, and so on. A few large carnivores are also shown—they were probably seen as rivals or enemies rather than as game. The relationship between these two categories in the famous painted cave of Lascaux in Dordogne, France, is typical: there are more than 200 figures showing game animals, but only 6 or 7 lions and 1 bear. Altogether, there are about 100 bear representations in Paleolithic art.

Moreover, when we look in detail at the bear pictures, it seems that most of them probably represent the living species *Ursus arctos*—the brown bear—and not the cave bear at all. To be sure, it is not easy to tell which species is meant, when they are so closely related as the brown bear and the cave bear, and moreover we do not know exactly what the cave bear looked like in life. In addition, we have no guarantee that the cave artists were concerned with exact realism (on the contrary, there is even one case of what seems to be a bear with the tail of a wolf).

One of the finest bear pictures comes from the cave of Teyjat in Dordogne. The animal certainly looks very like a brown bear. Although the head is well rounded, the limbs

29. Paleolithic engraving of a bear, probably *Ursus arctos,* from the cave of Teyjat in Dordogne, France. After Koby.

are relatively long and slim. The same species probably is represented by a very peculiar engraving in the cave of Trois-Frères in Ariège, France. This bear, according to Count Bégouen and the Abbé Breuil, seems to be vomiting its blood, and there are various signs on its body, some of them perhaps representing spears or other projectiles. The bear's flat and low head profile apparently proclaim it a brown bear and not a cave bear.

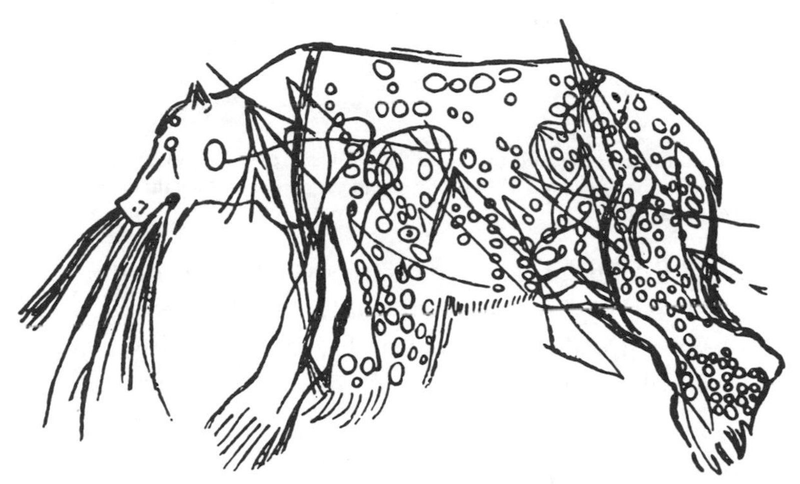

30. Another engraving, probably of the brown bear (*Ursus arctos*), from Trois Frères cave, in Ariège, France. It has been regarded as a bear wounded by spears and vomiting blood. After Koby.

Similar characters are seen in various other bear pictures, such as a painting in black pigment from a cave by Santimamiñe near Santander in northern Spain; two profile heads, one from Lascaux and one from La Madeleine, Dordogne; and a figurine from the Isturits Cave in the Pyrenees. There is no reason to regard any of these as anything else than brown bear.

Two loose slabs from the cave La Colombière in Ain, France, have engravings showing bears. One of them shows only the head, which has a rounded profile and an almost piglike snout. This may be a cave bear, as Abel suggests, but the evidence is hardly conclusive. The other slab shows the entire animal; the head is rather similar, but the limbs are fairly long and slender. Another creature of about the same type was depicted on a rock slab from Massat in Ariège. The chances are that all of these, too, are brown bears.

31. Head of a bear, engraved on a slab found in the cave La Colombière, in Ain, France. The shape of the muzzle and forehead suggested to Abel that it might be a cave bear. After Abel.

Finally, there is a remarkable engraving from the Combarelles Cave in Dordogne. It shows a very stocky, heavily built bear with short, powerful limbs, a hanging, vaulted head—all of these being characters that apparently distin-

32. Engraving from Les Combarelles, in Dordogne, France, possibly showing a cave bear (*Ursus spelaeus*). After Koby.

guished the cave bear. The bear's snout is well developed and does not exhibit the pug-dog type ascribed by some students to this species, but, as was noted in chapter 1, we do not have to assume such a trait. And so the bear from Les Combarelles may, perhaps, be an eyewitness portrait of the extinct cave bear. On the other hand, we can not be absolutely sure that it does not represent a very large, fat, brown bear!

The Les Combarelles bear is shown moving slowly ahead, or possibly lying dead on its right side. There are curving lines over the body which may, or may not, represent spears. The cave paintings were long interpreted as works of so-called sympathetic magic: by drawing an animal, especially one with a spear in it, a hunter gained influence over the real animal and ensured a successful hunt. It is still being done. You take a photograph of some one you hate, stick needles in it, and expect the victim to die.

But, as has been pointed out by Peter J. Ucko and Andrée Rosenfeld, for example, this is only one of numerous possible interpretations of cave art, and there is little reason to prefer it, especially since very few animals are actually shown wounded or in association with spears and the like. Alexander Marshack has found that many of the cave engravings were remade numerous times, apparently by different people. A ritual is indeed suggested, but its meaning is still unknown.

Probably the most remarkable art object in this connection is a headless clay sculpture of a bear found in 1923 by the intrepid speleologist Norbert Casteret in the cave of Montespan in the French Pyrenees. This is a life-size model, some two feet high (0.6 meters) and almost four feet long (1.2 meters) representing a massively proportioned bear, lying down on its belly. It is thought to have been originally covered by the skin of a bear, with the head fixed in its proper place by a wooden stick. The sculpture is riddled by spear

33. Headless clay sculpture of a bear from the Montespan cave in the French Pyrenees. A bear skull found nearby has unfortunately been lost. Redrawn from a photo by M. Bouillon.

marks, so it presumably was used for a ritual, perhaps of the sympathetic-magic type. M. Casteret and his assistant Henri Godin found the skull of a young bear between the forepaws of the sculpture. They knew that examination of the skull by a specialist would reveal which species was the object of this ritual.

In a letter of August 17, 1974, M. Casteret tells of the fate of this skull, discovered more than fifty years earlier. He left it in place to be viewed by the experts (the Abbé Breuil, Dr. Capitan, Count Bégouen, and Miss Garrod) who were immediately summoned to the Montespan cave. In the intervening two days, a small channel was dug to drain the inner part of the cave, which was flooded. But on returning to the statue, M. Casteret and the invited experts were startled to find the skull gone—stolen! So this skull, seen only by Casteret and Godin, was lost to science, and we shall probably never know which kind of bear was involved in the ritual of Montespan.

Although there is no confirmation of a cave bear cult, at least we may assume that the species was well known to early men—Neandertal men and, after them, Cro-Magnon men—who lived at the same time. We know that the brown bear and the grizzly bear have been assiduously hunted in modern times, even to their extermination in many areas. Did early man hunt the cave bear too?

The idea that there were tribes specialized in the hunting of cave bears, and that they were responsible for at least some of the accumulations of bear remains in caves, crops up from time to time. Professor Lothar F. Zotz even speaks of a bear-hunting phase in the economy of early man. An amibitious attempt to characterize such man-made assemblages was made by Heinz Bächler, the son of Emil Bächler. On the basis of a careful analysis of isolated teeth, he was able to show certain differences in the age structures of the bear populations of different caves. In some caves, the number of cubs and young animals was especially high, and these he interpreted as bear-hunting stations; for, no doubt, early man would have found the immature bears easier to kill than the adult.

There are several reasons to reject the suggestion of specialized bear-hunting tribes. In the first place, the high phosphate content of the bear cave earth proves that many of the animals were left to rot on the spot and were not eaten. Phosphate is also formed in caves settled by man, but the content is much lower. As to the large number of young found in most caves, this is only to be expected from natural mortality, which strikes most heavily at the immature and the aged (as was indeed pointed out by Professor Elisabeth Schmid); we return to this in chapter 7.

Then, there are very few, if any, stone implements in most bear caves. Any prolonged settlement by Paleolithic man tends to be marked by the sprinkling of innumerable flint flakes. The skinning and cutting up of a killed bear is quite an undertaking, and in the process more than one flint implement is likely to be damaged and discarded. There should also be butchering cuts on the bones. But the marks that are actually found, are either haphazard breaks due evidently to "dry transport" and the like, or marks left by scavengers and gnawing animals. Broken-up long bones have been thought to show that man broke them to get the marrow out, but the long bones of a bear do not have an easily extracted marrow like those of an ox or sheep. The hyena can utilize the nutriment concealed in the bone of a bear, by smashing it and eating it as such, but this is beyond man.

In the typical bear cave, there are also lots of shed milk teeth, in which the roots have been resorbed. This proves that young bears were hibernating peacefully in the cave at the time, for such teeth came from living bears, not from dead.

Many caves also show other mementos of the presence of bears. Scratch marks and footprints occasionally occur, but of course do not necessarily prove that the cave was visited more than a few times. The so-called *Bärenschliffe* tell a very different story. They are found in narrow passages, on the ceiling or the walls, and sometimes on loose slabs that are now found imbedded in the cave earth but which once formed part of the ceiling or wall. They are surfaces polished to a mirrorlike sheen by the passage of innumerable bears during hundreds or thousands of years. Few things speak more eloquently of the vastness of geological time and the cumulative numbers of living beings that have trodden one and the same path than these bear-polished limestones.

The big bones of the cave bear may well have been useful to man as implements. A thighbone would have made a splendid club, and a mandible with the canine tooth remaining could have been used as a scraping or digging tool. But there is little evidence of the actual use of cave bear bones. The wear marks found on the cave bear teeth are natural ones, brought about by the bear itself when it was still alive. Sometimes, pieces of bone show a polish of the same type as the *Bärenschliffe* and very likely due to similar causes: partly imbedded in matrix, the exposed part of the bone was worn by animals swishing by.

Some of the big canine teeth of the cave bear actually wore down in such a peculiar manner that unwary students have been led completely astray. One type of wear tends finally to weaken the tooth so much that its outer part breaks off and is lost. That discarded tooth fragment looks so much like a knife blade, well polished by use, that it was once described as the "Kiskevély knife" (after a Hungarian site of that name), supposedly of human manufacture. It was Koby who presented the true explanation of this odd pseudoimplement.

In stories about the Ice Age, the cave bear is generally depicted as a comparatively easy prey, in spite of its great size. In contrast, the brown bear is thought to have been much respected and avoided. There are various restorations and accounts of the supposed methods of cave bear hunting.

Perhaps the most vivid account is that given by Othenio Abel, who speculated about the manner in which Ice Age man might have hunted the bear of the Mixnitz Dragon Cave. He suggested that the hunters might have found it advantageous to arrange an ambush inside the cave, while the bear was away. When the bear finally appeared, it was killed or stunned by a rapid hit over the nose. The important thing (averred Bachofen-Echt) was to damage certain nerves, thus producing instant paralysis. As a medical man, Koby immediately branded this a tall story: damage to the olfactory nerves, which are the only important nerves in this area, will produce neither paralysis nor instant death.

The hunter being right-handed, the ensuing damage to the skull of the clubbed bear should be found on the left side. Other accounts of cave bear hunting also stress damage to the left side of the skulls of wounded bears. However, of the skulls from the Mixnitz cave, six are damaged on the left side, sixteen on both sides, and one on the right; incidentally, all of the skulls probably were damaged after death. Two skulls show partially healed lesions that were caused during life, but whether they were caused by a weapon wielded by man, by rockfall from the ceiling of the cave, or by some other agency, would be difficult to decide.

A celebrated find from the Sloup Cave in Moravia, Czechoslovakia, was published by Jindrich Wankel in 1892. Wankel found the top part of a skull (probably, but not certainly, that of a cave bear), with a partially healed lesion. Some hours after the finding of the skull, two workmen discovered a flint piece in the same part of the cave. Could this have been part of the weapon that caused the lesion and then stuck in the head of the bear, finally to be dislodged after the bear had died and its flesh disintegrated? Unfortunately, the flint piece does not look much like any sort of projectile point, least of all like the Solutrean laurel point shown in Wankel's figure reproduced here (figure 34).

It is not uncommon to find a bear skull with peculiar lesions on top of the head. Are we to assume that Neandertal man when wishing to kill a bear, habitually took a swipe at the top of its head? Let us not underestimate the intelligence and professional knowledge of these early men, who lived by hunting and assuredly knew all there was to know about the effects of their arms. You can kill a man by hitting him over the head, but to kill a bear that way, and particularly a cave bear with its immense sinus cavities, would call for more than superhuman strength. In fact, some of the lesions found on cave bear skulls are probably due to rockfall, while others point to inflammations with

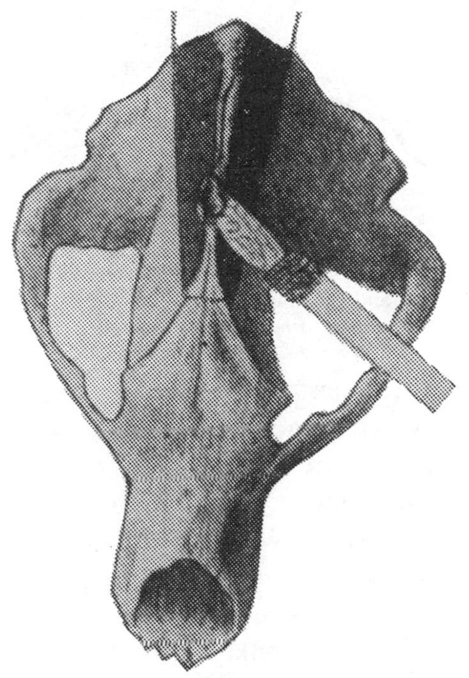

34. Wankel's sketch of a cave bear skull. The skull is thought to have been damaged by a projectile point in the manner shown; actually, only the calotte fragment (darker color) was found. Below is a flint piece found nearby (not to the same scale). Sloup Cave, Moravia.

resulting osteolysis—the bone is "eaten away"—as will be discussed in chapter 7.

So the various accounts showing holes in the tops of bear skulls and the attempts to fit flint weapons or bone clubs into these holes, seem somewhat futile. Neandertal man, who did not know the bow and arrow, would perforce

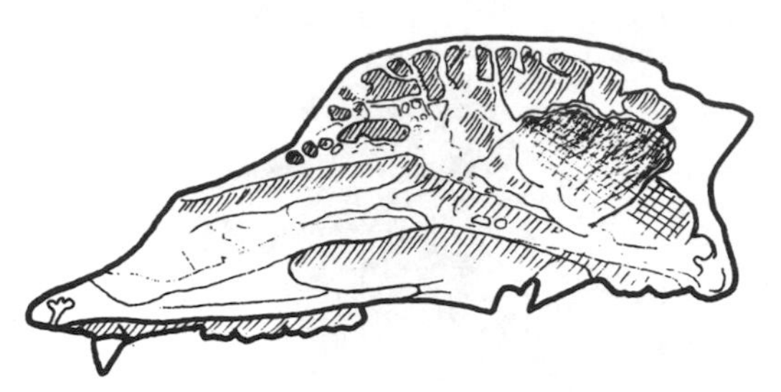

35. Sectioned cave bear skull, showing the nasal cavity, the large air sinuses in the upper part of the skull, and the comparatively small braincase well down in the hind part of the skull. Redrawn after Koby and Schaefer.

have to choose between the bludgeon and the spear. Not much deliberation was needed to make the right choice when a bear hunt was in the offing.

There are, of course, other methods of hunting that may have been known to Neandertal man. Camouflaged pit traps were probably used to catch large game, which could then be killed with spears or by throwing rocks. Such traps are, however, difficult to make in the hilly or mountainous regions inhabited by the cave bear, and we have found no evidence of them there.

Broadly, three types of caves containing cave bear remains can be distinguished. The first is the exclusive bear cave, like most of those discussed here—the Drachenloch, the Wildenmannisloch, the Mixnitz cave, and so on—where most or all fossils are remains of the cave bear, and only a few traces of man are found. Such caves are known from the Pyrenees, the Alps, and further east into the Caucasus.

A second group consists of the caves that were settled intermittently by man and by bears in the intervals between human occupancy. There are numerous examples of these, too, a good one being the Akhshtyrskaya Cave in the Caucasus, on the right bank of the Mzymta River. Located about 330 feet (one hundred meters) above the present-day river bed, it was intermittently inhabited by man over several millennia, beginning with early Mousterian times and going on to historical times. But there were also long intervals when the cave was forgotten by man and was used by bears and bats. There are many caves of this type, for instance the Veternica Cave not far from the city of Zagreb in Croatia, Yugoslavia, and the classical cave of Gailenreuth by Muggendorf in Franconia, Germany.

The third type of bear cave is the true hunter station, in which most or all of the bones present were brought in by man. Cave bear bones may be found in these caves too, but they are very rare and are completely overshadowed by the bones of characteristic game animals. Again, the history of the Caucasus as set forth by the paleontologist N. K. Vereshchagin, gives us a typical example of the hunter-station cave in the Sakazhia Cave in western Transcaucasia. This cave was inhabited by Upper Paleolithic men who left behind thousands of flint tools and fragments belonging to the Solutrean culture tradition. Among the animal bones found in the cave, those of bison predominate greatly; there are 1,488 bison bones, and they must represent at least 32 individuals. The number of cave bear bones, in contrast, is only 35, and no more than 5 individuals are represented. These bones may, perhaps, be spoils of the hunt and the same may be true for the remains of at least 3 brown bears found at the same site.

A site that does not quite fit into any one of these categories is Érd, in Hungary, which is an open-air Mousterian hunting station. As in typical bear caves, some 90 percent of the bones are from the cave bear. There are also horse, woolly rhino, and other game animals, but of the latter animals almost only skull and limb bones were found, while the bear skeletons are represented in their entirety. This indicates that the bears died, or were killed, on the spot, while the other animals had been killed elsewhere, and selected parts brought in. Probably this was a favored living site for man and bear alike, and perhaps one in which man was in fact hunting the bear. But it is a very isolated case.

We may now begin to suspect that there is something wrong with the popular picture of the cave bear as an easy prey and the brown bear as a difficult one. As a matter of fact, it seems to be the other way around. There are at least two instances known of large-scale bear hunting and possibly even ritual behavior by Neandertal men, but in both cases the species hunted was the brown bear, not the cave bear.

One of these instances was described by the French prehistorian Eugène Bonifay in the early 1960s, from excavations in the Régourdou Cave not far from the well-known Lascaux site in Dordogne. In connection with a Neandertal burial, there were various offerings made, including the upper arm bone of a brown bear. Other remains of the same species were found in positions suggesting intentional deposition. So we find the brown bear, but not the cave bear, associated with Neandertal rituals.

The other instances of bear hunting come from the interglacial travertines in the vicinity of the city of Weimar, Germany. In this area are numerous warm springs, whose mineral-laden waters deposit a calcareous tufa, the travertine. In the dry and cold climate of the glaciation, these springs dried up, but there is a long series of deposits from the interglacial, and they are quite rich in fossils. At Taubach, no less than twenty-two species of mammals are represented by fossil remains. They include such animals as the beaver, hamster, straight-tusked elephant, Merck's rhinoceros, wild ox, bison, giant deer, red deer, moose, roe deer, fallow deer, and wild boar. These animals evidently lived in a forest of temperate type. The site has also yielded remains of Neandertal man and various stone artefacts of Mousterian type. The eminent paleontologist Wolfgang

Soergel recognized long ago that the animal remains represented food refuse of Neandertal men.

The most common species at Taubach is the rhinoceros, which seems to have been the preferred prey. Carnivore fossils are quite scarce, with the single exception of the brown bear; the bones and teeth of this species must represent at least forty-three individuals (the real number probably rises into the hundreds).

With few exceptions, the bear fossils are fragmentary; for instance, not a single complete skull has been found. Such fragmentation, as we have noted, may well be due to natural causes. But there is one peculiar trait that is not found in the bear caves: the canine teeth have been systematically shattered (see figure 36).

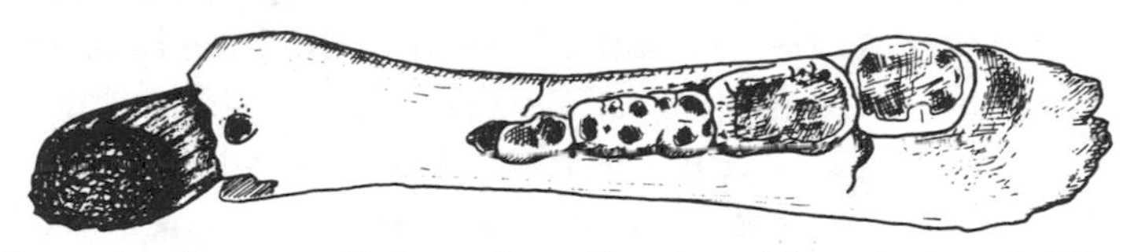

36. Lower jaw fragment of a brown bear (*Ursus arctos*) from the interglacial travertines of Taubach, German Democratic Republic. The cheek teeth show normal wear, but the canine tooth crown has been artificially destroyed.

The condition is not the result of normal wear. The teeth have apparently been deliberately smashed; only the roots are intact. This is true for about 80 percent of the canines. The cheek teeth, in contrast, show natural wear only. We do not know exactly how and why Neandertal man destroyed the big canines of the bear skulls and mandibles. The point is that the species involved is, again, the brown bear and not the cave bear. In fact, only four teeth of the latter species have been found at Taubach. At nearby Ehringsdorf, cave bear remains are more numerous, but they still do not equal those of the brown bear.

All this goes to show that, at least since interglacial times—80,000 years ago—man has been interested in the brown bear, the species that is still with us; interested enough to hunt it, to paint and engrave its likeness on the walls of his caves, and, at least in more recent times, to make it an object of cult and ritual. In antiquity, the bear was connected with the worship of Artemis, evidently because at an earlier time the goddess herself was a bear, and one of the constellations of the northern night sky is still called the Great Bear.

Whilst the brown bear moves in the limelight, his giant cousin the cave bear remains in the shade as far as man is concerned. Why this should be is not easy to understand. With its primarily vegetarian disposition, the cave bear has been thought to have been a comparatively docile animal. We may perhaps suspect that this idea is completely wrong. The African buffalo is also a complete vegetarian, but he does not impress one as docile. Probably the cave bear in reality was an extremely irascible creature, whose gigantic strength and thick skin made him almost invulnerable.

But that cannot be the whole story. The great brown bear is also a most redoubtable animal, and yet the Taubach Neandertal man pursued and destroyed it. Perhaps the solution to the problem lies in the habitats of the two species. Did the cave bear prefer areas which were difficult of access to man or areas lacking such attractions as good game and fishing? We have seen that few cave bear caves show evidence for more than transient and rare visits by human beings. Perhaps, after all, encounters between man and cave bear were relatively few.

NOTES

The reports on the Drachenloch are by E. Bächler (1921, 1940). The relationships between the cave bear and man were discussed by Koby (especially 1953b, but also earlier papers) whose reasoning has been followed, with some additions, in this chapter. On the Petershöhle see Hörmann (1923); on Mixnitz see Abel and Kyrle (1931), and on Wildenmannisloch, E. Bächler (1934). Bear burials by Lapps are described by

Zachrisson and Iregren (1974). Abel and Koppers (1933) studied bear representations in Paleolithic art. For an introduction to Paleolithic art see Ucko and Rosenfeld (1968). Many new facts and ideas were brought forward by Marshack (1972). Bear hunting by Paleolithic man has been speculated on by many authors, for instance Abel and Kyrle (1931); see also Soergel (1922) and Zotz (1951). The significance of age structure in samples of cave bear was interpreted by H. Bächler (1957), whose conclusions were rejected by Schmid (1959). Many examples of *Bärenschliffe* are found in Abel and Kyrle (1931). On the wounded cave bear from Sloup, see Wankel (1892) and Koby (1953b). The sites of the Caucasus are discussed in Vereshchagin (1959), those of Veternica in Malez (1963), and those of Érd in Gábori-Czánk (1968), Bonifay (1962) reports on Régourdou, and Kurtén (1975 and in press) on the bears of Ehringsdorf and Taubach. On bear rites in antiquity see Matheson (1942).

CHAPTER SEVEN

Life and Death

None of us will ever see a living cave bear, but there are dead cave bears galore. Now is the time to ask how, when, and why did they die?

In its area of life, the cave bear probably had few dangerous enemies. Man was one, but, as we have seen, only a few of the remains in the caves can be regarded as spoils of the hunt. What about the contemporary carnivores?

In his great monograph on the brown bear, the French naturalist Marcel A. J. Couturier names the wolf the bear's most dangerous enemy besides man. Wolf packs have been known to attack and bring down adult bears. There is also a record from the Ussuri basin in the USSR of tigers attacking brown bears, but this is quite exceptional. The tiger and the bear go out of each other's ways, and the same was probably true of the cave lion and the cave bear. Unprotected bear cubs are, of course, easy prey to these and other carnivores.

Probably a cave bear in its prime had little to fear from

contemporary carnivores, although a pack of cave hyenas may occasionally have overwhelmed a cave bear. In general it seems that the two animals moved in different spheres: where the hyena is common in caves, the cave bear is scarce, and vice versa.

Diseased bears may have been easier victims, but only if very enfeebled; a cave bear suffering from pain but with its strength little impaired must have been an exceedingly ugly customer and one to be avoided.

With its vegetarian habits, the cave bear probably steered clear of encounters with the fierce large ungulates of its day—the bison, aurochs, wild boar, rhino, and so on. Fights between bears may well have occurred, but in most cases fighting between members of the same species is more or less ritualized and fatalities kept to a minimum.

What remains, then? We may enumerate the causes of cave bear death as accidents, hunger, thirst, old age, and disease.

Accidents probably played an important part in cave bear mortality. Stonefall in caves has already been mentioned. Also, many caves, among them Gailenreuth, form natural traps; the bear seeking shelter falls down a pit and is dashed to death, or if it survives the fall, is unable to climb out. Most of the remains in Gailenreuth were found beneath a steep ledge where the bears had tumbled down. Very large caves with a maze of galleries may trap the unwary intruder, who gets lost in endless passages and starves to death.

But the remains in most caves are of bears that died during hibernation. As was shown by Professor Kurt Ehrenberg of Vienna, to whom we perhaps owe more of our knowledge of the cave bear than to any other scholar, the remains in the Mixnitz cave form a series of distinct growth stages: newborn (the size of a rat); one-year-olds (the size of a wolf); two-year-olds (the size of a hyena); three-year-olds (the size of a lion); and adults. Birth probably took place in the months of November to February, much as in the brown bear. Death in the cave would mainly come to those individuals who had not succeeded well in building up a store of fat during the summer season, and the cumulative hardships of wintering probably took their main toll near the end of the hibernation period when such stores were running low. Young individuals, still dependent on their mothers, succumbed if the mother died. At several places in the Mixnitz strata, remains of two or more newly born cubs were found together, often close to the spring in the cave; Ehrenberg suggested that they may have belonged to the same litter. They may have died of some contagious disease or because the mother died. Sometimes the remains of a one-year-old were found near the remains of such a litter. Ehrenberg pointed to the so-called "nurse" or *pestun* of the living brown bear in Russia—an older cub who still keeps company with its mother and the young of the year.

Ehrenberg also noted the strain imposed by tooth replacement in the young bears. The very small milk teeth were shed and replaced by the huge permanent teeth, still much too large for the small jaws of a second-winter cave bear. The amount of rebuilding of the jaws was amazing. Let us observe the gyrations performed by the last lower molar as it develops. When it forms at first, it is enclosed in the ascending ramus of the jaw, and stands almost vertically, front end pointing down and crown surface turned inward. From this position, still encased in the jawbone, it is slowly rotated into its mature position, while the jaw grows to accommodate it: the crown swings toward the horizontal, the front end lifts, and the hind end is pushed down until finally the molar takes its place in the tooth row of the two-year-old bear.

Some other teeth, notably the big canines, made almost as complicated evolutions as they grew into position. Even slight mechanical damage to the jaws could result in the process being disturbed, perhaps with a severe inflammation in train. Impaired chewing could lead to disease in the stomach and intestine.

This brings us to disease as a cause of death. In fact, the list of maladies diagnosed by the Austrian pathologist Richard Breuer, by Ehrenberg, and by many others, is a long one.

Many cave bears clearly suffered from osteoarthritis. This disease often produced arthritic outgrowths of bone, so-called exostoses, sometimes of fantastic appearance. Vertebrae or limb bones often fused together into a single mass of bone that must have made its bearer more or less lame.

Rickets was another common malady, which was related to the feeding habits of the bears and probably also to their long sojourns in dark caves without sunshine. As Ehrenberg noted, evidence of this disease is particularly common in high Alpine caves such as the Dachstein cave at 7,200 feet (2,200 meters above sea level); it is less common in the Mixnitz cave, 3,300 feet (1,000 meters above sea level); and almost unknown in the Winden bear cave at an elevation of only about 520 feet (160 meters). The length of hibernation, he suggested, is directly related to the altitude of the site: the higher the cave, the shorter the summer season.

Other infirmities seem to be due to the heavy use of various organs. Ehrenberg noted many cases of exostoses on the forearm bones due to inflammations in ruptured muscles, tendons, or periosteum (the lining of the bone); these cases show the entire scale from healthy specimens to severely damaged ones. Heavy wear of the teeth led to exposure of the roots and pulp cavities with resulting festering. Caries has been observed in some cave bear teeth.

Koby has noted that the great frontal sinuses of the cave bear skull were prone to infections resulting in osteolysis—an eating away of the bone that produces perforations; such perforations have irregularly rounded, smooth borders and so are easy to distinguish from lesions brought about by mechanical damage. Koby thought they might have resulted from an attack of parasitic worms, as is the case in ferrets and some other members of the weasel family.

Still another group of diseases arose from mechanical damage due to accidents, blows, bites, and the like. Healed fractures are often found among cave bear fossils, for instance broken limb bones that have reknit at odd angles; bears thus afflicted were permanently crippled. There are even several instances of the *os penis,* or baculum, having been broken and reknit. In bears, a fracture of the penis bone is not necessarily fatal for the urethra is not encased in the bone as it is in dogs.

Broken teeth, especially canines, are a common sight.

37. Left tibiae (shin bones) of cave bear. The pathological condition of the specimen to the left is due to a fracture of the head of the bone. Redrawn after a photo by Ehrenberg.

As I write this I have before me the skull of a young cave bear, probably no more than five or six years old, a magnificent fellow in its very prime (see figure 4, chapter 2). The right upper canine was broken during life and the bone around it has been eaten away by a severe inflammation. The animal probably was unable to eat normally and so ended its life during hibernation in an emaciated and feverish condition.

Some bears probably died, in the last analysis, from sheer old age; with teeth reduced to senile stumps, they were unable to masticate their food properly and so succumbed during hibernation.

In most of these cases the bear probably found its last resting place in a cave: the center and home of its territory, as we have some reason to believe. If we follow this line of thought, it becomes probable that many, perhaps most, of the cave bears actually died in caves, and so quite a large proportion of the cave bears that lived in Europe during the late Pleistocene may have wound up as fossils (mostly, of course, in fragmentary condition) and are in principle available to science. It is a unique situation. I do not know of any other species of which the same could be said. Here is indeed a challenge to the paleontologist.

What can we learn from this remarkable fossil record? Let us turn to one of the prolific cave bear sites and see what it can tell us about the life and death of this animal. The material to be described here was collected more than a century ago by the Finnish naturalist Alexander von Nordmann, then a high school teacher in the city of Odessa on the Black Sea. When he returned to his native Finland, he brought his vast fossil collection along, and it became the nucleus of the present-day Museum of the University Department of Geology and Paleontology in Helsinki.

Nordmann collected fossils in two caves, situated in Odessa and the neighboring village of Nerubay. His excellent study of the cave bear, published in 1858, is still a classic in its field.

As is the case with most old collections, we have no detailed stratigraphy of the caves, but a radiocarbon date processed on the collagen of cave bear bone indicates that these bears lived about 27,000 years ago. This takes us back to Hengelo-Denekamp times, the phase of temporary amelioration within the last glaciation during which men of modern type appeared in Europe. No trace of fossil man or his handiwork was reported by Nordmann, but twenty-three other species of mammals besides the bear have been identified in the collection. They are mainly "warm" or "temperate" forms such as the wild boar, red deer, fallow deer, roe deer, and corsac fox. There are also occasional northern animals such as the reindeer and the mammoth, suggesting that the sequence of strata continued into the last cold phase of the glaciation.

Still, as in all true bear caves, the overwhelming majority of the remains belong to *Ursus spelaeus.* The sex ratio is close to fifty-fifty, so the caves were about equally attractive to male and female bears. Most of the fossils are isolated teeth. An analysis of these, together with those teeth still sticking in jaws, gives us a picture of the dynamics of the bear population, of its balance between births and deaths.

To begin with, we must look at the milk teeth. They can be made to tell quite a story. It is useful here to focus on the milk canines, which are far more numerous than other milk teeth—there are in fact more than 200 of these little, pointed teeth, which seem so ridiculously small for an animal that would grow to the size of the great cave bear. Not all the milk canines are in the same stage of development. In the first place, there are quite a few that are no more than germs: they consist of nothing but an enamel cap, and the root has barely started to form. These, then, belonged to very young individuals in which even the milk

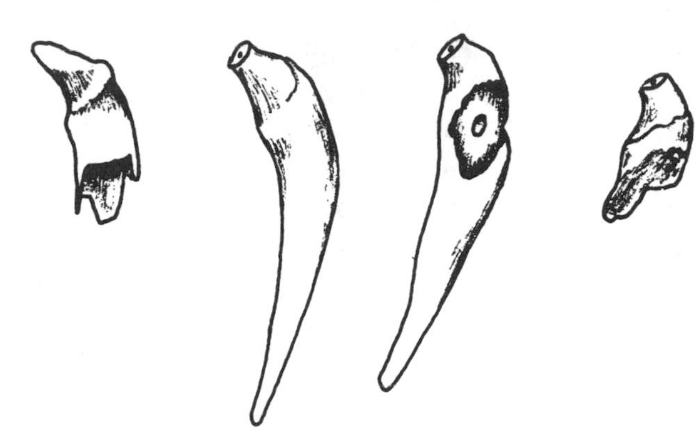

38. Milk canines of young cave bears from Odessa, USSR. Left to right: enamel cap of tooth still concealed in the jaw, root beginning to form; unworn tooth (tip broken) with fully formed root; tooth with root showing resorption marks; shed milk canine with resorbed root.

teeth had not yet been cut; in fact probably some of them are from unborn cubs that were still within the mother's body at her death and which died with her.

Next come teeth which have developed slightly further: the roots have formed, but the crown has not been worn. They must be the teeth of suckling cubs in their first winter, cubs which never knew the outside world, which lived on their mother's milk in the dark cave during the long months of late winter and spring but died before they were released by the warmth of summer.

And what comes next? Ah, this is interesting. Next comes, so to speak, nothing at all—for quite a while. The bears have left the cave, and there is a gap in the sequence. We lose sight of the cubs; when they return to spend their next winter in the cave, many things have happened. They have grown, and they are cutting their permanent teeth.

You can tell this from the milk canines. As a preliminary to the tooth being shed, its root is gradually dissolved, and there are various specimens in which you can see this process going on. The crowns are worn too, which indicates that the bears were using their teeth during the summer.

Some teeth show a very early stage of wear and resorption of the root; these are probably animals that died in late summer, and this may suggest that the cave was a "home" in summer too, especially for the sick and defenseless. But most show heavy wear of the tooth crown, and the roots are partially eaten away by resorption. These individuals died in their second winter, before the shedding of the teeth.

Finally, there is a large crop of milk canines in which the root has dissolved completely. And here our record of mortality stops, as far as the milk teeth are concerned. These teeth do not represent deaths; they are shed teeth, and the bear lived on after shedding them. The shedding took place during the second winter, probably not far from the turn of the year, although the time probably varied. Of course, some of the bears represented by these shed teeth may still have died later on that same winter, but we have to go to the permanent teeth to find a record of that.

Now let us see what the milk teeth tell us about the early mortality of the cave bear. In this collection, with a total of 204 milk canines, 13 are of unborn or newborn cubs and 26 of suckling cubs. So the total of first-winter mortality is 39 out of 204, which is 19.1 percent. This figure covers the time from just before birth, in November or December, up to the time when the cave was finally vacated in April or May. Roundly, it gives us the mortality for the first half year of life.

As far as second-winter mortality is concerned, we only get the story for part of the time—up to the shedding of the canine teeth. Here the deaths are 25 out of 165, or 15.2 percent. This is a minimum figure for the second winter and, as we shall see, the permanent teeth give a considerably higher estimate.

It is quite easy to recognize the permanent teeth of yearling cubs: they consist mainly of the enamel caps of the crowns, and their roots have formed only partially. But this situation varies in accordance with the sequence in which the teeth are cut. The premolars and first molars, which are among the first teeth to come into position, have almost fully formed roots; the hindmost molars, on the other hand, are in a very early stage of development (see figure 39). Out of a total of 1,946 cave bear teeth from the Odessa caves, 743 are in this stage, which gives a second-winter mortality rate of 38.2 percent. We must conclude that most of this mortality was concentrated to the later half of the hibernation, after the shedding of the milk canines, as might indeed be expected.

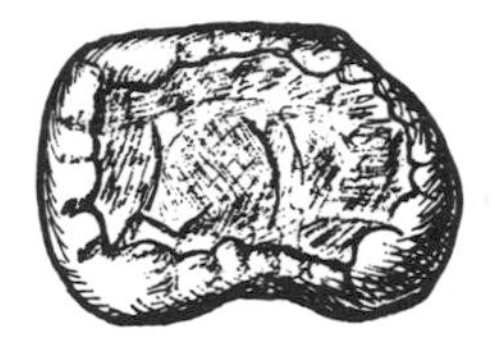

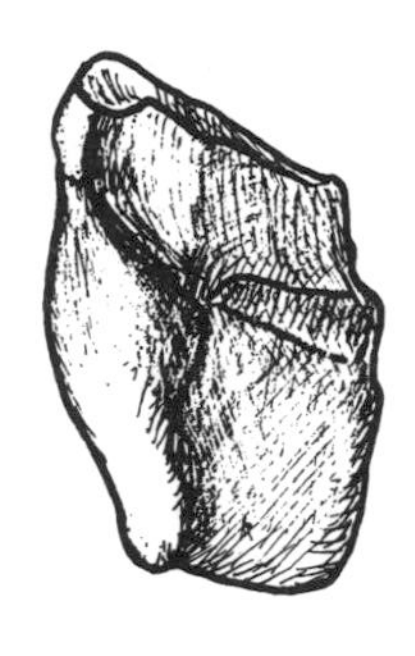

39. Third lower molars of a cave bear, Odessa caves, USSR. Left: crown and side views of the tooth of a one-year-old cub; the root has barely started to form and the tooth consists mainly of an enamel cap. Right: side view of the tooth of an old individual, where almost all of the crown has worn away and a wear facet has formed down along the side of the root.

The figure sounds startlingly high. It would mean that, out of every ten cubs that went into hibernation for their second winter, almost four, on average, died. Thus, it is not surprising that some students have been tempted to ascribe such a high degree of mortality to the influence of hunters who specialized in the capture of bear cubs for the table. But there is no sign of the presence of man. What is the situation in other caves?

To check the figures for Odessa, in the eastern part of the range of the cave bear, there are data for three other caves—two in Switzerland, in the center of the range, and one in Spain, at its western end.

One of the Swiss caves is Cotencher, not far from Neuchatel, and its fossil content was studied by the great naturalist H. G. Stehlin. Although Mousterian man left traces of his presence in the cave, Stehlin was convinced that all or almost all of the cave bear remains were accumulated by natural causes and represented bears dying peacefully in the cave. The bear stratum dates from the early half of the last glaciation.

Checking the numbers of yearling cubs by the same method as at Odessa, we find that they are represented by 131 teeth out of a total of 364. This makes a mortality rate of just 36 percent, a figure that is very close to that for the Russian cave.

The Spanish cave is Cueva del Toll, in the vicinity of the town of Moyá in Catalonia. Here the corresponding figures are 158 out of 419, or 38.6 percent—again, a closely similar result, and, again, an accumulation of bear remains without any indication that man had anything to do with them. The bear stratum of Cueva del Toll is evidently somewhat later in time than that from Cotencher and probably dates from about the same phase of the last glaciation as that from Odessa.

In all of these caves, the numbers of males and females are about equal. If the yearling cubs followed their mothers, as we have reason to believe, we ought to get different figures for caves in which the females outnumber the males. Such a cave, as we have noted in chapter 5, is Saint-Brais in the Jura, with 72 percent females and 28 percent males. And, in fact, the yearling teeth here are 144 out of a total of 302, or 47.7 percent. On the other hand, the number of two-year-olds is very low at Saint-Brais, suggesting that the young bears parted company with their mothers during the second summer. Caves with unequal representation of the sexes thus tend to give biased estimates of cub mortality. But we are recompensed for this by the information they give about the behavior of the bear.

In a male-dominated cave, like that at Mixnitz, it should be the other way around: relatively few one-year-olds, relatively many two-year-olds. This theory seems indeed to be confirmed by the collection at hand, although it should be remembered that only a small fraction of the original fossil content is now available, and it may be somewhat biased.

Returning now to the Odessa cave bear, similar estimates have been made for the later intervals in the life of the bear, and the results can be gathered into a so-called life table, of the type that insurance mathematicians make up for human populations. A life table for the cave bear is

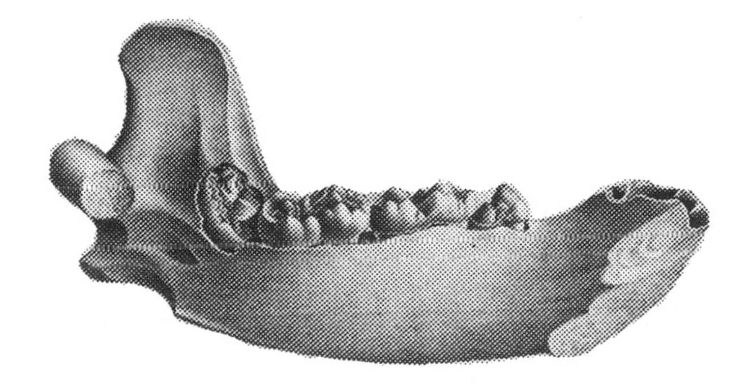

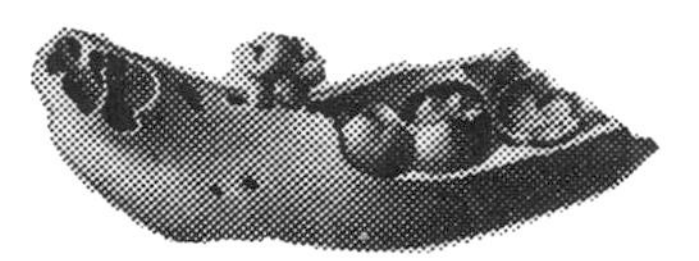

40. Lower jaws of cave bear cubs, Odessa caves, USSR. Bottom: young of the year, left jaw in outer view with milk molar in place and permanent teeth still concealed in jawbone. Top: one-year-old, left jaw in internal view with emerging permanent cheek teeth; note last molar still turned sideways in ascending branch of jaw. After von Nordmann.

given in the appendix to this book, together with a description of how to read it. At this point we only have to take note of what it tells us in a general way about the life history of the cave bear.

We can see that the mortality for cubs and young individuals is quite high but tends to become lower as the bear reaches adulthood at about four years of age. From an annual mortality rate of more than 35 percent, the figure is reduced to about 13 percent or so for adult bears in their prime. That is to say, a bear that had run successfully the hazards of cub mortality and attained adult age would have good prospects of survival for several years. But with advancing age the rate of mortality is again increased, and for bears of fifteen or more rises very high.

Exactly how old a cave bear could become is not yet known, but it is evident that its length of life was severely restricted by the heavy wear of its teeth (see figure 41). Brown bears in captivity may reach an age of thirty or

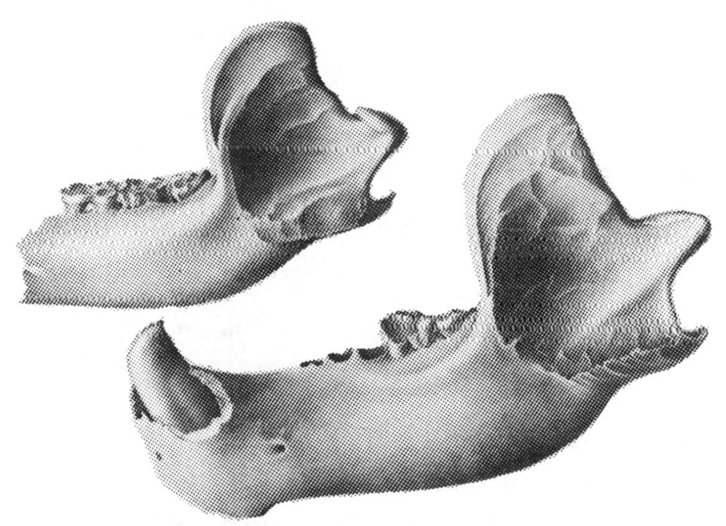

41. Left lower jaws of adult cave bears, outer views, Odessa caves, USSR (not to same scale as figure 40). Top: adult female with well-worn cheek teeth (front end of the jaw is broken off). Bottom: old male with remaining cheek teeth showing heavy wear extending to roots of penultimate molar. After von Nordmann.

more, but in the wild they apparently rarely live to more than twenty-five. By then their teeth are senile stumps. Those of the cave bear probably wore out more quickly, and I doubt that any cave bear lived beyond about twenty years of age.

A life table for the grizzly bear has recently been set up by John and Frank Craighead, who, with their collaborators, have studied the grizzlies of Yellowstone Park for more than thirteen years—by marking, radio-tracking, and other methods. Their results agree so closely with those for the Ice Age cave bear as to be almost identical (see appendix); but as the grizzly appears to be slightly longer-lived than the cave bear, the number of adults forms a somewhat higher proportion of the census than in the cave bear population.

The cave bear population had to be replenished each year by so many newborn cubs, and we can calculate that each adult female, on the average, had to produce slightly more than one cub a year to keep the numbers from shrinking. Again, a comparison with data on *Ursus arctos,* especially the Craighead observations on Yellowstone grizzlies, suggests that this is a realistic estimate. Actually, the litters found by Ehrenberg in the Mixnitz cave often consisted of two cubs.

Without delving further into statistics, we still have to note that the crude annual rate of mortality (that is, deaths at all ages, from newborn to old) was about 20 percent, or one bear out of every five. And this knowledge makes it possible to answer an interesting question: *was the cave bear a common or rare animal?*

When we look at the fabulous concentration of bear remains in some caves, our first impression must surely be that these animals were very numerous indeed. There are in existence old life restorations showing droves of bears, almost like a herd of buffalo or some other grazing animal (such a scene was reproduced in figure 2, chapter 1). The German geologist Albrecht Penck, one of the great pioneers in the study of the Ice Age, was even moved to say that Europe must have been flooded by cave bears at some time in the Pleistocene, most probably, he thought, during the Eemian interglacial.

But Professor Wolfgang Soergel, of Freiburg, Germany, soon pointed out that a different explanation was possible. Many of the caves, he stated, served as lairs for cave bears for thousands of years. The Mixnitz cave, for instance, was inhabited by bears both before and during the Weichselian glaciation; this gives a total duration of perhaps 100,000 years or even more. And we need only one bear dead in the cave every other year, on average, to add up to the estimated 30,000 to 50,000 individuals. Soergel concluded that the amassing of remains was due to intensive sampling out of a continuously small standing population.

Just how small? One bear dead every other year indicates a population of about 2.5 individuals or, as we cannot operate with fractional individuals, a population fluctuating in size between, say, one and four. Sometimes there might be a female with her young; at other times, two or three bachelors wintering in different parts of the cave, and so on. Far from being flooded by bears, the cave bear range was quite sparsely populated at all times. And so the natural history of the cave bear is taken out of the fantastic and returned to the realm of reality and credibility.

NOTES

On the brown bear see Couturier (1954). An excellent discussion of caves and fissures as fossil sites, with profiles of many caves including Gailenreuth, may be found in Zapfe (1954). Ehrenberg (1931) summarizes data on cave bear development and diseases; on the latter see also Abel and Kyrle (1931), Koby (1953a), and Tasnádi-Kubacska (1962). Nordmann (1858) describes the fossil bear from the Odessa caves; radiocarbon dating is discussed in Kurtén (1969b). The age grouping and life table treatment are emended from Kurtén (1958). On Cotencher see Dubois and Stehlin (1933); on Cueva del Toll see Donner and Kurtén (1958); on Saint-Brais see Koby (1938). Principles of life table treatment can be found in Deevey (1947); dynamics of the living grizzly population is treated in Craighead et al. (1974). Soergel (1940) discusses the causes of mass occurrences of the cave bear.

CHAPTER EIGHT

The Substitute Cave Bears

"But it's all *arctos!*"

I still remember my surprise when, years ago, I started to study the British cave bears. I knew, of course, that brown and grizzly bears (now regarded as variants of the same species, *Ursus arctos*) had been reported from British caves, together with *Ursus spelaeus.* In 1846 Richard Owen stated, "With the *Ursus spelaeus* was associated another bear, more like the common European species, but larger than the present individuals of the *Ursus arctos.*" In the continental bear caves, it is not uncommon to find an occasional specimen of brown bear among the great mass of cave bear bones. This was the situation I expected to find in the British caves, but it turned out to be completely different.

Tornewton Cave, in South Devon, is an example of a bear cave excavated by modern methods; it was dug during many years by Dr. Antony J. Sutcliffe and yielded a long sequence of fossiliferous strata, ranging in age from the Saalian glaciation to the end of the Pleistocene and into modern times. In Saalian times the cave was inhabited by bears, which have left their remains in great numbers in the lower strata. With the onset of a milder climatic regime in Eemian times, the cave became a hyena den, and the bear is found no more, except for a few scraps. Weichsel-glacial occupation was more sporadic, with moose and red deer during interstadial conditions and mainly reindeer during the intensely cold end phase of the glaciation.

The Saalian bears of Tornewton Cave behaved much like the continental bears. There are many remains of cubs in the cave, and there is the telltale age grouping suggesting that the cave was used for hibernation in winter but rarely if at all during the summer. All is according to pattern, except for one thing.

It is the wrong species. It is not the cave bear at all. Every bone and tooth in Tornewton Cave belongs to the brown bear, *Ursus arctos.*

The situation is the same in most other bear caves in Britain. For instance, the famous Victoria Cave near Settle in Yorkshire was excavated a century ago by the British Association for the Advancement of Science, and yielded a great fossil fauna of Eemian interglacial date. Both species, *Ursus arctos* and *Ursus spelaeus,* were reported as present there; but a restudy of the material showed that only the former species occurs; the "cave bear" was simply a huge brown bear. In fact I know of no evidence for the presence of *Ursus spelaeus* in Britain during Saalian and Eemian times.

In cave faunas from the last glaciation, the brown bear is still the predominant species, and reports of the cave bear are mostly mistaken. Still, there are at least two exceptions. The Weichselian deposits of Kent's Cavern in Torquay carry not only brown bear remains, but also cave bear remains in quite respectable numbers (and the basal strata of the cave have yielded a small, ancestral cave bear of much earlier date). Wookey Hole, a famous cave in Somerset, also has true cave bear remains—again, of Weichselian date. So it seems that the great continental species did range into southern England at some time in the course of the Weichselian glaciation. Yet so firm was the position of the brown bear in its British stronghold that even at these sites it outnumbers the cave bear.

The brown bear, then, acted as a substitute (or, to use the technical term, a vicar) of the cave bear in Britain. This is unexpected. Students of the living *Ursus arctos* stress that this species avoids caves as winter quarters. Dr. Peter Krott, who lived for several years in the Alps with the bears he had brought up ("they are not really dangerous, as long as you behave in a bearish way yourself," he comments), noted that they dug their own dens, although natural caves were available in the area. The Craighead brothers, speaking on the basis of unparalleled experience, emphasize the same for the grizzlies of Yellowstone Park. And yet we find the British bears violating the code of brown bear behavior and denning in caves!

Behavior is plastic and evolves in adaptation to environment and mode of life. After all, the brown and cave bears were closely related, and the absence of the other species may have made it possible for the brown bear to step into the "niche" of the cave bear on British soil.

How close the parallel was is a matter of speculation. It may be that in this area, a less exclusively vegetarian form was better equipped to make a living. Whatever the reason, it is a fact that the British Isles came to be a stronghold of *Ursus arctos* throughout the later Pleistocene, and it survived well into historical times. In 1781 W. Pennant noted that bears survived in the mountains of Scotland to as late as the year 1057. Ireland also had a Pleistocene and postglacial population of bears.

Through most of the historical range of the species in

the Old World, fossils of the brown bear may occur in caves, but apart from Britain they are never very common. Such sporadic finds as there are come from North Africa, Europe, and Asia. At the Peking man site of Choukoutien in China, some skulls and bones were found. These represent a very large type of brown bear, and some of them are so immense that Dr. Pei Wen-chung thought the bear might be a true cave bear. But the anatomical details agree better with the brown bear, so I think this bear must be *Ursus arctos* too (see figure 42).

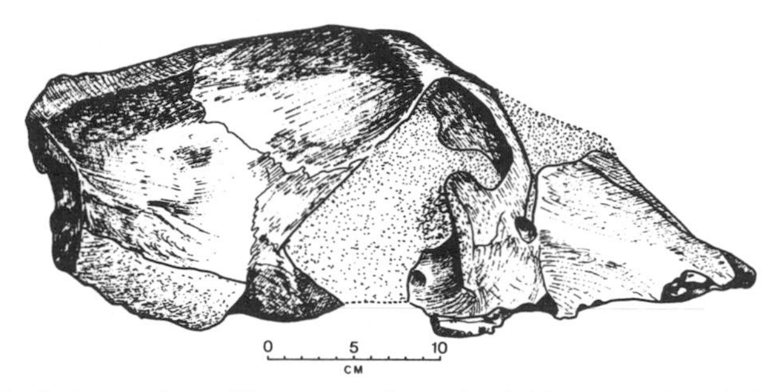

42. Skull of a brown bear, *Ursus arctos,* from the Peking Man site of Choukoutien, China. The enormous size of the skull led to its erroneous identification as that of a cave bear. After Pei.

In North America, except Alaska, the grizzly bear—which is the local form of the species *Ursus arctos*—is a rather late immigrant. During the last glaciation, the way south from Alaska was barred by an enormous ice field extending all the way from ocean to ocean. Towards the end of the glaciation, the ice melted and an ice-free corridor was formed through which animals (including man and the bear) could migrate south. But although the grizzly appears at various open-air sites in North America, including the famous tar pits of Rancho La Brea in Los Angeles, California, it is not present in caves.

The black bears, to judge from their fossil record, are somewhat more prone to cave-denning than the brown and grizzly bears. Much of the evidence comes from the so-called "dragons' teeth" or *Lung che* that used to form a very popular line of merchandise in Chinese drugstores. They were taken as remedies for most illnesses under the sun, but especially for heart and kidney disorders, and in cases of epilepsy; to say nothing of failing potency, an affliction where imagination tends to work wonders. As we have seen, the same practice was common in early European medicine. Professor G. H. R. von Koenigswald, a noted paleontologist and student of fossil man, tells in his reminiscences how he used to go the rounds of the Chinese drugstores of the East. Always there were dragon teeth first class (*Fun lung che,* big white dragon teeth) and second class (*Tsing lung che,* small black dragon teeth). The first-class teeth were heavily fossilized and dated back to the Tertiary or even older times; fossils from the Miocene *Hipparion* fauna are particularly common. The second-class teeth were less heavily mineralized, and from their preservation it is a safe bet that most of them came out of Pleistocene caves. (Wily businessmen often mixed in a few modern bones, suitably blackened with soot to look old; they are of course readily recognized by an expert. A time-honored test is to try the bone with the tip of the tongue: if it feels sticky, it is not fossil.)

A common species among such second-class stuff is the Tibetan black bear, *Ursus thibetanus.* This species is also quite common at the Peking man site of Choukoutien and in other Chinese caves (where it often masquerades under thc scicntific namc *Ursus angustidens,* thc "sharp-toothcd bear"). The living Tibetan black bear is a nocturnal animal, which spends the daytime in caves, thickets, and even hollow trees; such are also used as winter dens.

The American black bear, *Ursus americanus,* resembles its Asiatic cousin in its selection of hibernation places. Fossils of this species are often found in caves, and there may even be mass occurrences that almost bring the European bear cave to mind. The great collection from Cumberland Cave, in Maryland, represents at least thirty or forty individuals, but the actual number probably was much greater. Bears of all age groups are found there, showing that the animals actually inhabited the cave. Other American caves with large numbers of black bear fossils are Conard Fissure near Buffalo River, Arkansas, and Potter Creek Cave in Shasta County, California. The last-mentioned dates from the last glaciation, while the Cumberland and Conard bears lived in mid-Pleistocene times.

Chinese drugstore collections of "second-class" dragon bones and teeth may also include the remains of the small Malay bear, *Helarctos malayanus,* indicating that this species too, occasionally inhabited caves. As in the case of *Ursus arctos* and the two black bear species, the Pleistocene animals were markedly larger than those of the present day.

As to the other bear of southern Asia, the aswail or sloth bear *Melursus ursinus,* just one fossil specimen has been found in a cave; it comes from one of the Karnul Caves in Madras, India.

The Andean bear of South America, *Tremarctos ornatus,* is the last survivor of a great tribe of bears that ranged widely through the Americas in Pleistocene times. A closely related species was the Florida cave bear, *Tremarctos floridanus,* whose remains have been found in Mexico and the southern United States—California, New Mexico, Texas, Tennessee, Georgia, and especially Florida. Although many of the finds come from caves, there are no mass occurrences like those of the European cave bear, or even the Cumberland black bears, and so the name might seem ill chosen. But there is a point to it. The bodily resemblance of this American species to the European cave bear is almost uncanny.

Of course there are differences. Anatomical details

make it clear that the Florida cave bear was closely related to the living Andean bear, and their connection with the *Ursus* bears is certainly rather distant. Yet evolution, working with such different raw materials, brought forth a creature mirroring the European form in some of its most conspicuous features (see figure 43).

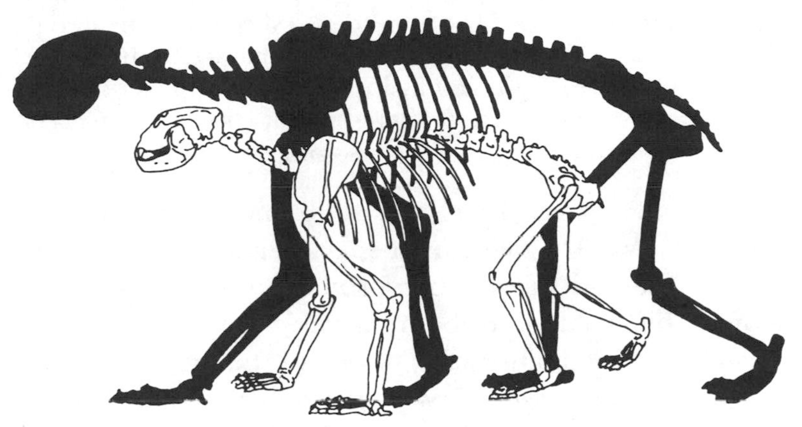

43. Reconstructed skeletons of a female (white) and a male (black silhouette) Florida cave bear, *Tremarctos floridanus.* The sexual dimorphism in size is somewhat enhanced by the fact that the female is smaller than average, while the male is a particularly large specimen. After Kurtén.

The Florida cave bear was a big animal. The weight of a large male has been estimated at some 650 pounds (upwards of 300 kilograms), while the much smaller female weighed about half as much. It was very heavily built, with a barrellike rib cage, short, broad paws, and elongated upper arm and thigh bones. The anterior premolars were reduced, the back teeth enlarged, and the jaw articulation shifted well above the plane of the teeth. The profile of the forehead shows a distinct step. The neck was lengthened, the back sloping, and the hindquarters were relatively weak.

All this could just as well read as a list of the special features in which the cave bear, *Ursus spelaeus,* differed from its close relative the brown bear. It seems clear that, within limits, the Florida cave bear was trying to do just the same as its European counterpart.

> The convergence between *Tremarctos floridanus* and *Ursus spelaeus* suggests that special adaptation to glacial conditions can have played only a minor role in the evolution of the distinctive characters of the cave bear. Here we see a species living in the equable climate of the Gulf Coast, far from the ice sheets and the periglacial, and yet acquiring most of those same characters. It may well be asked what factors of habit and environment, of heritage and development, resulted in bringing about this amazing convergence in surroundings that seem so dissimilar.

To these lines, which I wrote in 1966, might be added that it probably was the vegetarian specialization, plus reliance on great size and strength for protection, that was the key factor in the evolution of both species.

This is a good illustration of one of the modes of evolution—that of convergence: two originally dissimilar animals come to resemble each other more and more, by adapting to similar ways of life. Convergence is classically exemplified by whales and dolphins, which have evolved to resemble fish in adapting to life in the sea. Of course, convergence always implies a still earlier phase of divergence. Very long ago, for instance, *Tremarctos* and *Ursus* had a common ancestor—probably at the *Ursavus* or *Protursus* stage in the Miocene—from which they diverged. In the same way, mammals and fish had a common ancestor in the still more remote past, from which they diverged when mammalian ancestors emerged onto the land and fish ancestors remained in the sea.

The Florida cave bear survived to the end of the Pleistocene and possibly into the first millennia of postglacial time. In Florida it has been found associated with early man.

North America was also inhabited by more distant relatives of the Andean bears, but they are regarded as members of another genus, *Arctodus.* Estimates of the weights of the largest of these bears exceed 1,300 pounds (almost 600 kilograms), indicating that they were gigantic animals that would dwarf even the Kodiak bear and the great cave bear. But in contrast to cave bears, they were rangy, long-limbed beasts, probably quite fast-moving animals in spite of their huge size, and evidently rapacious carnivores rather than vegetarians.

Some of these bears seem to have made caves their home too, for their remains are found in caves in California, Texas, Missouri, Maryland, and Pennsylvania. Mostly, however, the great *Arctodus* bears are found at open-air sites, and in such circumstances they range from Mexico to Alaska. In South America, related species are found, both in caves and in the soils of the Argentinian pampas. Still, none of the *Arctodus* bears is particularly common as a cave dweller, and so we may pass them over here.

Thus, when all is said, the true cave bear, *Ursus spelaeus,* remains unique.

NOTES

Owen's (1846) work is the classic on British fossil mammals and birds. The sequence at Tornewton Cave is described by Sutcliffe and Zeuner (1962). A series of annual reports by Tiddeman (e.g. 1876) records the excavations in Victoria Cave. Much of the information given here on British fossil bears is based on unpublished work by myself, but see e.g. Kurtén (1959, 1968, 1969a). Pennant (1781) records the survival of brown bears in Scotland in historical times. The carnivores of Choukoutien are described by Pei (1934), the postglacial grizzly from Rancho La Brea by Kurtén (1960). On Chinese drugstore teeth see Koenigswald (1955) and Erdbrink (1953, especially pp. 53–55). The fossils from Cumberland Cave are discussed in Gidley and Gazin (1938). On the Florida cave bear see Kurtén (1966), and on the *Arctodus* bears see Kurtén (1967a).

CHAPTER NINE

The Extinction

Most of the species that ever existed on earth are now extinct. Some died out because their "niche" vanished—that combination of environmental factors and adaptive response that kept the species going. Others disappeared because their niche was conquered by a superior competitor. Species have vanished as victims of predation, as a result of catastrophes, and, lately, from deliberate extermination by man.

In the still-influential theory of the extinction of the cave bear, which was developed by Othenio Abel many years ago, none of these factors is invoked. As outlined already in chapter five, this theory holds that the species became extinct because of degeneration, or what we would perhaps now call inadaptive genetic drift.

According to this theory, the cave bear lived for a long time under optimal conditions, without serious enemies. It was thought that in such a situation natural selection would be much relaxed, and thus many kinds of inferior variants were able to survive and reproduce. In the end the population became saturated with degenerative strains, and was so much weakened that it could easily be extinguished—for instance, by the change in climate at the end of the last glaciation.

This theory seeks an explanation of extinction in the established interaction between genes and selection and so is one step ahead of many less creditable ideas that were much in vogue during the years between World War I and World War II. Just as an example, we may note the idea of "racial senility," according to which a species could become old and decrepit in the same way as a single organism. The idea is surely meaningless; a species is born anew with every generation, and so the analogy is false. As far as we know, life arose only once on earth, and in that sense all living things are exactly the same age.

Loss of adaptation due to genetic drift, on the other hand, is a distinct possibility. As symptoms of degeneration, Abel pointed to the appearance of stunted individuals, the so-called dwarfs (especially at Mixnitz), to the supposed differential birthrate with an excess of males, to the high variability in general, and to the large numbers of diseased individuals.

But the Mixnitz dwarfs are not dwarfs, they are normal females; there was no differential birthrate, more than in other bears, for instance the grizzly; the variation in local cave bear populations is by no means particularly great, compared with that in other bears; and the frequency of diseased individuals is a perfectly natural phenomenon. This last conclusion is reached by the following reasoning: the cave bears we see are dead; they must have died of something; disease would have been one of the important factors of death; ergo, we must expect to find many diseased individuals. The living bears probably were quite healthy. But the material that has come down to us has been selected, by death itself, for such failings as immaturity, disease, and old age.

Soergel, who in the main accepted Abel's degeneration theory, thought that the cave bear was more vulnerable than the brown bear on a number of scores. First, the productivity of the cave bear was thought to be lower than that of the brown bear; that is to say, there were fewer cubs in the litter. Second, he thought that the rate of mortality was higher in the cave bear than in the brown bear. Third, the potential length of life for the vegetarian cave bear, overstraining its dentition, would be shorter. Fourth, Soergel accepted the assumption of a differential birthrate. Although the first and third assumptions may be right, the second probably is not (rates of mortality seem to be much the same in cave and grizzly bears), and the fourth is clearly wrong.

Is there any reason, then, to assume that the cave bear was a victim to genetic drift and resulting loss of adaptation? It was apparently assumed that a life without dangerous enemies would be a nonselective situation, which would lead to degeneration. But the cave bear is by no means the only species that has existed for untold millennia without having any serious predators at its heels. Many large herbivores, such as the elephants and rhinos, are so big and powerful that no carnivore in its right mind dares to attack them; yet they do not show degenerative traits. Predation by enemies is by no means the only selective factor in the life of an animal species, and it is not even the most important one in most cases.

Selection has been defined as the agency that produces a systematic shift in the gene pool of a species over the generations, such a systematic shift being what we call evolution. There is another kind of evolution too: the unsystematic or random evolution that may occur in small populations where chance rather than selection determines which genes are passed on to the next generation. This process is called genetic drift, and it results in gene loss—the genetic variation becomes impoverished rather than increased.

It is, of course, possible that some isolated local cave bear populations were so small that genetic drift became important, but if so we have little evidence of it, except perhaps towards the very end of the existence of the species. On the other hand, there are some aspects of natural selection that may well be studied with fossil material.

Although selection makes its impress on the population in many different ways, in the final analysis they all work out as differential fertility: certain genetical traits tend to be favored, at the expense of others, from generation to generation. One such mode of selection—but, we should keep in mind, only one out of many—is differential mortality: the varying degree of success of differently endowed individuals in the "struggle for existence," to use Alfred Russell Wallace's famous phrase. This is precisely the mode that can be investigated with fossil remains such as those of the cave bear, where natural mortality at all stages of life is abundantly represented. If we find that those who died young tended to differ systematically in some traits from those who survived to adult and old age, then we have evidence of differential mortality; or, in other words, evidence that natural selection did occur.

For such a study it is of course necessary to choose a trait that does not change with age. Fortunately, mammals have in their permanent teeth traits of just this kind. Once a tooth crown has been formed, it will remain unchanged except for being gradually worn by constant use. (Some teeth, in some mammals, have open roots and grow throughout life, and they cannot be used for this kind of study; but bear teeth do not come into this category.) Even if the wear is heavy as in the cave bears, there are many dimensions that will not be affected, or are affected only in a few senile specimens, and they form ideal material for the study of selection.

The results of such studies—which consist of statistical analyses of large numbers of teeth—have indeed shown that selection was at work. In general it can be said that individuals with teeth of a size and shape close to the average tend to be more common among those who survived to adult or old age, whereas more extreme variants are much more common among those who died early. This kind of selection, favoring the average or "normal" type, is termed stabilizing selection.

Stabilizing selection indicates that the species was well adapted (as regards the trait being studied). Most populations in nature are likely to be well enough adapted to their current mode of life, because they have presumably been following the same trade for a very long time, and so they will have stabilizing selection.

But there may also be directional selection. In this case, individuals differing from the population average will tend to be favored.

The lower third molar of the cave bear from Odessa shows evidence of such directional selection (see figure 39, chapter 7). This tooth is somewhat variable in outline. In some individuals it is rather short and broad, in others very elongated. On the whole, elongation of this tooth is a hallmark of the cave bear; in no other bear species does the third molar reach such a remarkable length. Still, as noted, it varies.

Now it turns out that the teeth of those cave bears that died young—in their first three years—average noticeably shorter and also somewhat broader than those of bears that survived to adult age. Indeed, in the quite old individuals these teeth are particularly long and narrow. Now, there is no possibility for the shape to change during life of an individual, so the difference can only result from differential mortality.

The directional selection in this case favors a long, narrow tooth. During the history of the cave bear, the third lower molar gradually became more and more elongated, so it would not be too surprising to find that this type of selection was still at work in the Late Pleistocene, at least in some local populations. (It is not found in all populations—for instance, not at Mixnitz, where the selection was the stabilizing type, favoring the average shape.)

Of course the selection may also have had something to do with the complicated movements of this special tooth while it was still concealed in the jaw (a process discussed in chapter 7). Perhaps a long and narrow tooth went through these evolutions more lissomely than a short and broad one, and so was less liable to cause trouble.

Another kind of directional selection observed in the Odessa bear affected the size of a cusp on one of the molars, and it is thought that in some individuals this cusp did not mesh well with its receptacle in chewing (the receptacle was a "valley" in the opposing molar). Apparently, the cusp was sometimes slightly too big, or the valley too small, and so we find that a big cusp and a small valley are particularly common in those who died young, while individuals with a small cusp and/or a big valley tended to survive (see figure 44).

It has sometimes been stated that tooth cusps wear down so rapidly that they can have but the slightest selective importance. But those early times in life when the cusps wear down are also the times when mortality, in a wild mammal population, is particularly great. Also, it is at this time that the animals have to eat, not just to keep going, but to sustain growth. So it can be concluded that cusp patterns must have great selective value; and we can see this conclusion verified in the case of the cave bear.

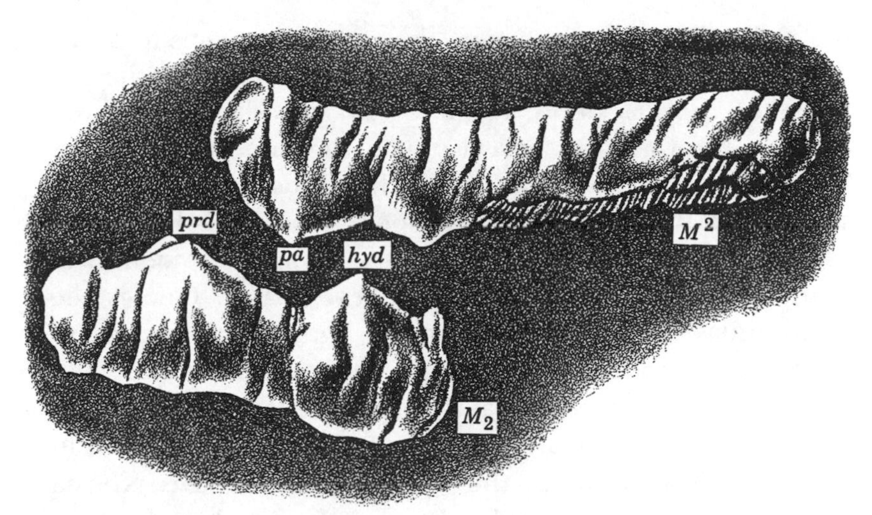

44. Occlusion between second upper (M^2) and lower (M_2) molars in a cave bear. The cusp at the front end of the upper molar, termed the paracone (pa), fits into the notch between the lower molar cusps, termed protoconid (prd) and hypoconid (hyd). The view shows the left side teeth from the outside. After Kurtén.

We have come a bit ahead since 1942, when Sir Julian Huxley (in his wholly admirable classic, *Evolution: The Modern Synthesis*) stated that paleontology, by its very nature, cannot throw any important light on selection. We can now hope that study of differential mortality, combined with observations on actual evolutionary trends through time, may give important new information to the student of evolution.

Shelving the theory of death by degeneration, let us return to the problem of the vanishing cave bear. What is the anatomy of its extinction?

In most bear caves, the evidence indicates that the species vanished well before the end of the Weichselian glaciation. On the other hand, there are also instances of very late survival. For example, a tabulation of Swiss caves by Karl Hescheler and Emil Kuhn lists fifteen sites with *Ursus spelaeus;* but only two of these contain evidence of occupation after the Mousterian—the era of the Neandertal man. These caves are the Schlossfelsen at Thierstein, and possibly the Kohlerhöhle; at both sites, the cave bear seems to be associated with Magdalenian man, and so dates from the last phase of the glaciation. A survey by Vereshchagin of the caves in the Caucasus reveals much the same information. There are six or seven caves with the Mousterian cave bear; seven with the cave bear in unspecified Upper Paleolithic associations; one Magdalenian cave (uncertain identification); and two caves that may be even later. One is the Gvardzhilas Cave near the village of Rgani; the bones there are human meal remains, and goat bones predominate, but there are also six cave bear remains (at least two individuals); the age is very late, at the end of the Pleistocene or in the early Postglacial. In the Vorontsovskaya Cave, Khosta Ravine, a few cave bear bones were found that show no fossilization; they may also be early postglacial.

In central Europe, too, the cave bear seems to have survived locally to the very end of the Ice Age and perhaps beyond. In the Moravian bear cave Pod hradem the record of the cave bear goes up to the uppermost Weichselian strata (see figure 45). In the Swabian Alps in Germany, the late cave bear is associated with Magdalenian industry. The two Westphalian sites, Balver Höhle and Hohler Stein, also have very late records of the species, and cave bear association with a Mesolithic industry is suggested in the former cave. The same may be true for the Tischofer Cave at Kufstein in the Tyrol, Austria.

It is thus clear that the extinction of the cave bear was not a sudden thing but a gradual process, spread out over several thousand years. Local populations would become extinct, beginning well back in the mid-Weichselian. The once continuous range of the species would become more and more patchy. Local populations would gradually become completely isolated from each other. The record shows that these local populations had somewhat reduced variability, perhaps due to gene loss from moderate inbreeding.

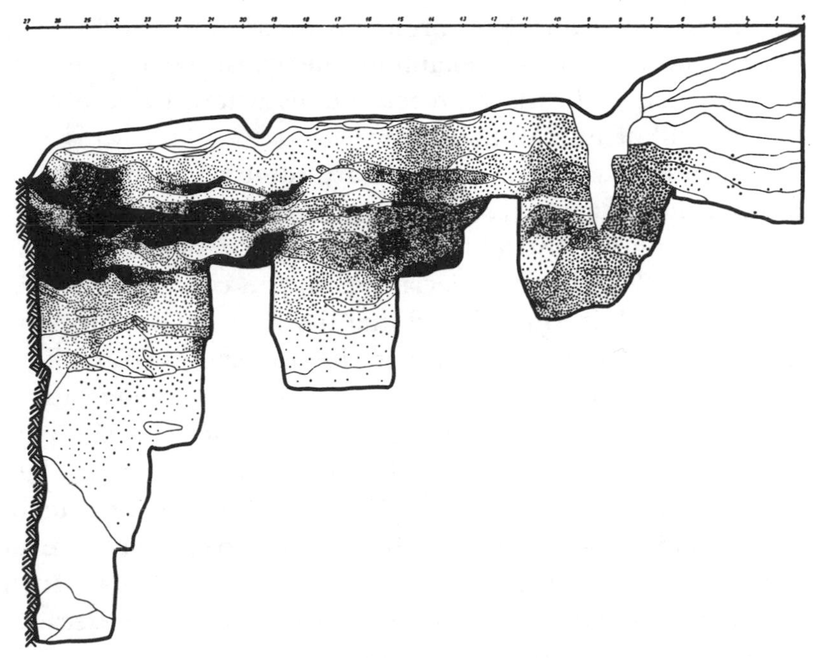

45. Profile of deposits in the Pod hradem bear cave, Moravia, Czechoslovakia, with the distribution of cave bear bones in the cave earth; each dot represents one specimen. Cave bear bones were found at all levels in the Pleistocene deposits of the cave; only those deposits formed after the Ice Age (upper-most layers) lack the cave bear. After Musil.

Small isolates of this type, if continuously kept at a low population level, are quite liable to become extinct by sheer chance. Once the bear was in this situation, extinction would be practically certain unless its living conditions were drastically improved. Today, we try to save endangered species by setting aside reservations, by breeding stocks in zoos, and the like. No such measures, of course, were taken on behalf of the cave bear.

There are some caves in which the evidence suggests that man succeeded the bear as an inhabitant; we have noted examples of this, for instance, in the caves of the Caucasus. So it might be thought that competition for caves between man and bear played some part in the gradual shrinking of the bear population. As Kazimierz Kowalski has noted, however, there are many areas in which this cannot be shown, and, for example, the Polish bear caves are situated mainly in the high mountains where there are no traces at all of human habitation.

Whatever the reason, there are some who think that a distinct increase of environmental pressure may be sensed in the changing distributions of fossil bear remains, at the end of the Weichselian glaciation. In the cave of Pod hradem, Dr. Rudolf Musil found that the number of immature individuals tended to increase in the uppermost strata. This cave, situated high up on the mountain wall beneath a ruined castle, overlooking the Punkva River, in Moravia, was excavated by Musil and Dr. Karel Valoch in the late 1950s; it is a typical bear cave. Musil grouped the bear teeth into juvenile and adult, and found that the percentage of juveniles increased in the late Weichselian deposits. Much the same had been noted by Professor Ehrenberg decades earlier in the Belgian cave Hastière.

This evidence may mean, as Musil suggests, that the dynamics of the population was actually affected, with increasing juvenile mortality. On the other hand, it might just mean that the caves were becoming more popular among females with small cubs, and that bachelors were being evicted. Indeed, Musil states that remains of smaller bears, presumably female, tend to be more common in the late deposits.

In Dr. Kowalski's opinion, the extinction of the cave bear was due to the vanishing of its biotope—that special environment in which the species existed for so long. Towards the end of the Weichselian, the tundras, subarctic taigas, and steppes of central Europe vanished, and the land was clothed once more in a temperate forest like that of the Eemian.

The problem is obvious: the end of the Weichselian was by no means the first time that such a transition had occurred. The cave bear, and its predecessors, had survived a long series of interglacials, some of which (the Eemian, for instance) were distinctly warmer than the interglacial in which we now live. Why, then, should they be wiped out this time?

The history of extinction suggests an answer. Many local populations had already become extinct well before the change from glacial to postglacial times, and the bear was already what we would now call an endangered species. The climatic changes at the end of the Ice Age might, then, have killed the stragglers.

If this was so, however, Kowalski's theory really begs the question, for the extinction, as we have noted, has its roots far back in Weichselian times. In fact, all the available data suggest that the decline of the cave bear started at about the time when Neandertal man vanished and modern types of men made Europe their home.

The problem is really a broader one, for the cave bear was not the only species to disappear at the end of the Ice Age. There was a wholesale extinction: in Europe, for instance, mammoths, rhinos, steppe bison, giant deer, musk oxen, leopards, and cave lions died out. The phenomenon was in fact world wide, and the animals affected were mostly large ones—from man's point of view, important game or dangerous predators.

It is true that cases can often be made out for a real loss of biotope. The present-day tundra of the high Arctic, for example, with its long polar night, is certainly different from the steppe-tundra of central Europe that was the homeland of the woolly mammoth and woolly rhino. But the same was presumably true for the Eemian situation, which the mammoths and rhinos took in their stride. And so the problem of the Pleistocene extinctions remains unsolved.

It is not made easier by the fact that many large Pleistocene mammals remain with us to the present day—in Europe, for instance, the brown bear, the bison, the moose, and the red deer. All of these, including the bear, were hunted by man as early as the Eemian. Thus, human predation in the Ice Age did not necessarily lead to extinction, and it is hard to see how it could, with such a very sparse human population. By and large, the Ice Age hunting peoples probably lived in harmony with their environment, harvesting the surplus rather than making inroads on the capital. In 1912, Wolfgang Soergel stated, "In all the hunting that ends with the extermination of a species, the motivation is never hunger. Money, and the greed for it, have been the incentive. The savage does not know these; he hunts to eat and so is unable to decimate the big game to any important extent."

So we are at an impasse. Perhaps we have to wait for a new way in which to view the problem; perhaps for a combination of factors, in which man and the changing climate fall into place as the kaleidoscope is turned.

NOTES

Abel's (1929) theory of the extinction of the cave bear was accepted by Soergel (1940). On various types of selection see Simpson (1954, in which also the notion of "racial senility" and other bizarre ideas are effectively dealt with). Differential mortality in cave bear teeth is discussed in Kurtén (1967 b); also in much unpublished material. Swiss fossil fauna is treated in Hescheler and Kuhn (1949), and Caucasus fossil fauna in Vereshchagin (1959). Further data on extinction can also be found in Kurtén (1958). The Pod hradem cave bear was studied by Musil (1965), the bears from Hastière by Ehrenberg (1935a). Many important studies on Pleistocene extinctions were brought together in Martin and Wright (1967); they include Kowalski's paper (cave bear, pp. 359–60) and also an excellent study by Vereshchagin. On Paleolithic hunting and extinction see also Soergel (1912).

APPENDIX

The Life Table

The life table (see Deevey, 1947) summarizes the fate of a "cohort" of individuals who start life together. For regular intervals of age, it gives the number of deaths, the number of survivors, the rate of mortality, and the expectation of life (or mean remaining lifetime). These columns are headed x, d_x, l_x, q_x, and e_x, respectively (1000 q_x indicates that the rate of mortality is given on a per mil basis).

The life table for the cave bear is given with one-year intervals, except for the first interval which is only 0.5 year. It was constructed by calculating, for each age interval, the ratio $q_x = a/(a+b)$, in which a is the total number of teeth belonging to the given interval, and b the sum total of all older teeth. Original values of q_x for the intervals between 4.5 and 15.5 years were somewhat irregular and have been smoothed by the use of sliding means for three consecutive years. The expectation of life was only calculated for every fifth year. Age determinations over 5.5 years are approximate and preliminary. The table is an emended version of one published in Kurtén (1958).

The table shows that rates of mortality are high in the first few years of life, and that only about one cub in 4 survived to adult age. The rate of mortality is gradually reduced and tends to fluctuate somewhat below 15 percent annually in middle life.

After about 12 years of age the rates again increase as senility sets in (probably mainly due to wearing out of the dentition). The expectation of life at birth is only about 3.5 years but rises to over 5 years in the young adult, then gradually to diminish with increasing age.

The age structure of the cave bear population may be compared with that of the Yellowstone grizzly bear, for which Craighead et al (1974) have constructed a 9-year average. The comparison is made in the second table. As may be noted, the sets of figures for the two populations run closely parallel to each other, but the relative number of adults is lower in the cave bear population. This is probably, in part at least, a real difference, and due to a difference in the potential longevity of the two species. Of course, it may also reflect a certain bias in the cave bear sample from Odessa. Such a bias could arise if adult skulls and jaws were used for exchange or as gifts.

Life Table for the Cave Bear (*Ursus spelaeus*) Population from Odessa

x *Age Interval*	d_x *Deaths During Interval*	l_x *Living at Beginning of Interval*	$1000q_x$ *Rate of Mortality*	e_x *Expectation of Life*
0–0.5	191	1000	191	3.47
0.5–1.5	309	809	382	
1.5–2.5	113	500	227	
2.5–3.5	76	387	198	
3.5–4.5	59	311	189	
4.5–5.5	39	252	155	5.1
5.5–6.5	30	213	141	
6.5–7.5	25	183	136	
7.5–8.5	21	158	134	
8.5–9.5	18	137	131	
9.5–10.5	18	119	151	4.2
10.5–11.5	16	101	158	
11.5–12.5	18	85	212	
12.5–13.5	18	67	269	
13.5–14.5	17	49	347	
14.5–15.5	13	32	407	1.5
15.5–16.5	9	19	474	
16.5–17.5	6	10	600	
17.5–18.5	4	4	1000	

Age structures of cave bear (*Ursus spelaeus,* Odessa) and grizzly bear (*Ursus arctos,* Yellowstone Park) *

Percentages of:	*Cave Bear*	*Grizzly Bear*
Cubs	23.5	18.6
Yearlings	14.5	13.0
2-year-olds	11.3	10.2
3-4-year-olds	16.4	14.7
Adults	34.2	43.7

* Data from Craighead et al., 1974.

Bibliography

Abel, O. 1929. *Paläobiologie und Stammesgeschichte.* Jena.

Abel, O., and Koppers, W. 1933. Eiszeitliche Bärendarstellungen und Bärenkulte in paläobiologischer und prähistorisch-ethnologischer Beleuchtung. *Palaeobiologica* 5:7–64.

Abel, O., and Kyrle, G., eds. 1931. *Die Drachenhöhle bei Mixnitz.* Speläolog. Monogr. Vols. 7–8. Vienna.

Arambourg, C. 1933. Révision des ours fossiles de l'Afrique du Nord. *Ann. Mus. Hist. Nat. Marseille* 25:247–301.

Bächler, E. 1921. Das Drachenloch bei Vättis im Taminatal. *Jahrb. St.-Gall. Naturf. Ges.* 1920–21.

—— 1934. Das Wildenmannisloch am Selun (Churfirsten).

—— 1940. *Das alpine Paläolithikum der Schweiz.* Basel.

Bächler, H. 1957. Die Altersgliederung der Höhlenbärenreste im Wildkirchli, Wildenmannisloch und Drachenloch. *Quartär* 9:131–46.

Bonifay, E. 1962. Un ensemble rituel Moustérien à la grotte du Régourdou. *Sixth Proc. Int. Congr. Prehist. Rome* 1962:132–40.

Buckland, W. 1822. Account of an assemblage of fossil teeth and bones . . . discovered in a cave at Kirkdale, etc. *Phil. Trans. Roy. Soc.* 122:171–236.

Charlesworth, J. K. 1957. *The Quaternary Era.* 2 vols. London.

Colbert, E. H. 1962. The weights of dinosaurs. *American Mus. Novitates,* 2076:1–16.

Cooke, H. B. S. 1973. Pleistocene chronology: Long or short? *Quatern. Res.* 3:206–20.

Couturier, M. A. J. 1954. *L'Ours brun.* Grenoble.

Craighead, J. J., Varney, J. R., and Craighead, F. C. 1974. A population analysis of the Yellowstone grizzly bears. *Bull. Montana Forest & Cons. Exp. Station* 40:1–20.

Crusafont, M., and Kurtén, B. Bears and bear-dogs from the Vallesian of Spain. *Acta zool. Fennica.,* in press.

Cuvier, G. 1823. *Recherches sur les oseemens fossiles etc.* Nouv. ed., 4. Paris.

Deevey, E. S. 1947. Life tables for natural populations of animals. *Quart. Rev. Biol.* 22:283–314.

Dehm, R. 1950. Die Raubtiere aus dem Mittel-Miocän (Burdigalium) von Wintershof-West bei Eichstätt in Bayern. *Abh. Bayer. Akad. Wiss.,* n.ser. 58:1–141.

Donner, J. J., and Kurtén, B. 1958. The floral and faunal succession of "Cueva del Toll," Spain. *Eiszeitalter u. Gegenwart* 9:72–82.

Dubois, A., and Stehlin, H. G. 1933. La grotte de Cotencher, station Moustérienne. *Mém. Soc. Paléont. Suisse* 52:1–178.

Ehrenberg, K. 1931. Der Höhlenbär. *Aus der Heimat* 44:65–80.

—— 1935a. Die Pleistozaenen Baeren Belgiens. I. Die Baeren von Hastière. *Mém. Mus. Roy. Hist. Nat. Belg.* 64:1–126.

—— 1935b. Die Plistozaenen Baeren Belgiens. II. Die Baeren vom "Trou du Sureau" (Montaigle). *Mém. Mus. Roy. Hist. Nat. Belg.* 71:1–97.

—— 1942. Berichte über Ausgrabungen in der Salzofenhöhle im Toten Gebirge. II. Untersuchungen über umfassendere Skelettfunde als Beitrag zur Frage der Form- und Grössenverschiedenheiten zwischen Braunbär und Höhlenbär. *Palaeobiologica* 7:531–666.

—— 1964. Ein Jungbärenskelett und andere Höhlenbärenreste aus der Bärenhöhle im Hartlesgraben bei Hieflau (Steiermark). *Ann. Nat. Hist. Mus. Wien* 67:189–252.

—— 1966. Die Plistozänen Bären Belgiens. III. Cavernes de Montaigle (Schluss), Cavernes de Walzin, Caverne de Freyr, Cavernes de Pont-a-Lesse. *Mém. Inst. Roy. Sci. Nat. Belg.* 155:1–74.

Erdbrink, D. P. 1953. *A Review of Fossil and Recent Bears of the Old World.* 2 vols. Deventer.

Esper, J. F. 1774. *Ausführliche Nachricht von neuentdeckten Zoolithen unbekannter vierfüssiger Thiere. . . .* Nürnberg.

Flint, R. F. 1971. *Glacial and Quaternary Geology.* New York.

Gábori-Czank, V. 1968. La Station du Paléolithique moyen d'Erd-Hongrie. *Monum. Histor. Budapest* 3:1–277.

Geist, V. 1971. The relation of social evolution and dispersal in ungulates during the Pleistocene, with emphasis on the Old World deer and the genus *Bison. Quatern. Res.* 1:283–315.

Gidley, J. W., and Gazin, C. L. 1938. The Pleistocene vertebrate fauna from Cumberland Cave, Maryland. *Bull. U.S. Nat. Mus.* 171:1–99.

Heller, F. 1956. Thomas Grebners bisher unveröffentlichte "Descriptio antri subterranei prope Galgenreuth" aus dem Jahre 1748. *Geol. Bl. Nordost-Bayern* 6:32–40.

Hescheler, K., and Kuhn, E. 1949. Die Tierwelt. In *Urgeschichte der Schweiz,* ed. O. Tschumi, pp. 121–368. Frauenfeld.

Hörmann, K. 1923. Die Petershöhle bei Velden in Mittelfranken. *Abh. Naturhist. Ges. Nürnberg* 21:121–54.

Huxley, J. S. 1942. *Evolution: The Modern Synthesis.* London.

Koby, F. E. 1938. Une nouvelle station préhistorique, les cavernes de Saint-Brais. *Verh. Naturf. Ges. Basel* 49:138–96.

—— 1949. Le dimorphisme sexuel des canines d'*Ursus arctos* et d'*Ursus spelaeus. Rev. suisse Zool.* 56:675–87.

—— 1953a. Lésions pathologiques aux sinus frontaux d'un ours des cavernes. *Eclogae Geol. Helv.* 46:295–97.

—— 1953b. Les paléolithiques ont-ils chassé l'ours des cavernes? *Actes Soc. Jurass. Emul.* 1954:1–48.

Koby, F. E., and Schaefer, H. 1961. Der Höhlenbär. *Veröff. Nat. Hist. Mus. Basel* 2:1–25.

Koenigswald, G. H. R. von. 1955. *Begegnungen mit dem Vormenschen.* Düsseldorf.

Kozhamkulova, B. S. 1974. Zoogeographical analysis of theriofauna of Kazakhstan. *Trans. Int. Theriol. Congr. Moscow* 1:300–301.

Kurtén, B. 1955. Sex dimorphism and size trends in the cave bear, *Ursus spelaeus* Rosenmüller and Heinroth. *Acta Zool. Fennica* 90:1–48.

—— 1958. Life and death of the Pleistocene cave bear, a study in paleoecology. *Acta Zool. Fennica* 95:1–59.

—— 1959. On the bears of the Holsteinian interglacial. *Stockholm Contr. Geol.* 2:73–102.

—— 1960. A skull of the grizzly bear (*Ursus arctos* L.) from Pit 10, Rancho La Brea. *Contr. Sci. Los Angeles County Mus.* 39:1–7.

—— 1964. The evolution of the polar bear, *Ursus maritimus* Phipps. *Acta Zool. Fennica* 108:1–26.

—— 1966. Pleistocene bears of North America. I. Genus *Tremarctos,* spectacled bears. *Acta Zool. Fennica* 115:1–96.

—— 1967a. Pleistocene bears of North America. II. Genus *Arctodus,* short-faced bears. *Acta Zool. Fennica* 117:1–60.

—— 1967b. Some quantitative approaches to dental microevolution. *Jour. Dental Res.* 46:817–28.

—— 1968. *Pleistocene Mammals of Europe.* London.

—— 1969a. Cave bears. *Studies Speleol.* 2:13, 24.

—— 1969b. A radiocarbon date for the cave bear remains (*Ursus spelaeus*) from Odessa. *Comment. Biol. Soc. Sci. Fennica* 31 (6):1–3.

—— 1971. *The Age of Mammals.* London.

—— 1972. *The Ice Age.* New York.

—— 1975. Fossile Reste von Hyänen und Bären (Carnivora) aus den Travertinen von Weimar-Ehringsdorf. *Abh. Z. Geol. Inst.* 23:465–84.

—— In press. *Fossile Reste von Bären und Hyänen (Carnivora) aus den Travertinen von Taubach.* Weimar.

Lyell, C. 1875. *The Principles of Geology.* 2 vols. London.

Malez, M. 1963. Kvartarna fauna pećine Veternice u Medvednici (Die Quartäre Fauna der Höhle Veternica (Medvednica–Kroatien)). *Palaeont. Jugoslavica* 5:1–197.

Marshack, A. 1972. Cognitive aspects of Upper Paleolithic engraving. *Current Anthropol.* 13:445–77.

Martin, P. S., and Wright, H. E. 1967. *Pleistocene Extinctions: The Search for a Cause.* New Haven.

Matheson, C. 1942. Man and bear in Europe. *Antiquity* 1942:151–59.

Mottl, M. 1933. Zur Morphologie der Höhlenbärenschädel aus der Igric-Höhle. *Jahrb. Ung. Geol. Reichsanst.* 29:187–246.

—— 1964. Bärenphylogenese in Südost-Österreich. *Mitteil. Mus. Bergbau Landesmus. "Joanneum"* 26:1–55.

Musil, R. 1965. Die Bärenhöhle Pod hradem. Die Entwicklung der Höhlenbären im letzten Glazial. *Anthropos* 18:7–92.

Nordmann, A. von. 1858. *Palaeontologie Suedrusslands. I. Ursus spelaeus (odessanus).* Helsingfors.

Ovey, C. D., ed. 1964. *The Swanscombe Skull: A Survey of Research on a Pleistocene Site.* London.

Owen, R. 1846. *A History of British Fossil Mammals and Birds.* London.

Pei, W. C. 1934. On the Carnivora from Locality 1 of Choukoutien. *Palaeont. Sinica* C 8 (1):1–166.

Pennant, W. 1781. *History of Quadrupeds.* Vol. 2. London.

Rosenmüller, J. C. 1795. *Beiträge zur Geschichte und nähern Kenntniss fossiler Knochen. Erstes Stück.* Leipzig.

Rosenmüller, J. C., and Heinroth, J. C. 1794. *Quaedam de Ossibus Fossilibus Animalis cuiusdam, Historiam eius et Cognitionem accuratiorem illustrantia.* Leipzig.

Schmerling, P. C. 1833. *Recherches sur les ossemens fossiles découverts dans les cavernes de la province de Liège.* Vol. 1. Liège.

Schmid, E. 1959. Zur Altersstaffelung von Säugetierresten und der Frage paläolithischer Jagdbeute. *Eiszeitalter u. Gegenwart* 10:118–22.

Simpson, G. G. 1954. *The Major Features of Evolution.* New York.

Soergel, W. 1912. *Das Aussterben diluvialer Säugetiere und die Jagd des diluvialen Menschen.* Jena.

—— 1922. *Die Jagd der Vorzeit.* Jena.

—— 1940. *Die Massenvorkommen des Höhlenbären.* Jena.

Solecki, R. S. 1971. *Shanidar: The First Flower People.* New York.

Sutcliffe, A. J., and Zeuner, F. E. 1962. Excavations in the Torbryan caves, Devonshire. I. Tornewton Cave. *Proc. Devon Archaeol. Explor. Soc.* 5:127–45.

Tasnádi-Kubacska, A. 1962. *Paläopathologie.* Jena.

Tiddeman, R. H. 1876. Third report of the Victoria-Cave Exploration Committee. *Rept. Brit. Assoc.* 1875:166–75.

Toepfer, V. 1963. *Tierwelt des Eiszeitalters.* Leipzig.

Ucko, P. J., and Rosenfeld, A. 1967. *Palaeolithic Cave Art.* London.

Vereshchagin, N. K. 1959. *The Mammals of the Caucasus: A History of the Evolution of the Fauna.* Translated by A. Lermaw and B. Rabinovich. Jerusalem.

Wankel, J. 1892. *Die praehistorische Jagd in Mähren.* Olmütz.

Wolff, B. 1938–1941. Fauna fossilis cavernarum, I–III. *Fossilium Catalogus, Animalia.* Vols. 82, 89, 92, pp. 1–288, 1–320.

Woldstedt, P. 1969. *Quartär.* Stuttgart.

Zachrisson, I., and Iregren, E. 1974. Lappish bear graves in northern Sweden. *Early Norrland* 5:1–113.

Zapfe, H. 1954. Beiträge zur Erklärung der Entstehung von Knochenlagerstätten in Karstspalten und Höhlen. *Geologie* 12:1–59.

Zeuner, F. E. 1959. *The Pleistocene Period.* 2d ed. London.

Zotz, L. 1951. *Altsteinzeitkunde Mitteleuropas.* Stuttgart.

Index